Lecture Notes in Physics

T0213143

The Lecture Notes in Physics

The series Lecture Notes in Physics (LNP), founded in 1969, reports new developments in physics research and teaching – quickly and informally, but with a high quality and the explicit aim to summarize and communicate current knowledge in an accessible way. Books published in this series are conceived as bridging material between advanced graduate textbooks and the forefront of research and to serve three purposes:

- to be a compact and modern up-to-date source of reference on a well-defined topic

- to serve as an accessible introduction to the field to postgraduate students and nonspecialist researchers from related areas

- to be a source of advanced teaching material for specialized seminars, courses and schools

Both monographs and multi-author volumes will be considered for publication. Edited volumes should, however, consist of a very limited number of contributions only. Proceedings will not be considered for LNP.

Volumes published in LNP are disseminated both in print and in electronic formats, the electronic archive being available at springerlink.com. The series content is indexed, abstracted and referenced by many abstracting and information services, bibliographic networks, subscription agencies, library networks, and consortia.

Proposals should be sent to a member of the Editorial Board, or directly to the managing editor at Springer:

Christian Caron
Springer Heidelberg
Physics Editorial Department I
Tiergartenstrasse 17
69121 Heidelberg / Germany
christian.caron@springer.com

J. Reichardt

Structure in Complex Networks

Jörg Reichardt
Univ. Würzburg
Inst. Theoretische Physik
Am Hubland
97074 Würzburg
Germany
reichardt@physik.uni-wuerzburg.de

Reichardt, J., *Structure in Complex Networks*, Lect. Notes Phys. 766 (Springer, Berlin Heidelberg 2009), DOI 10.1007/978-3-540-87833-9

ISBN: 978-3-642-09965-6   e-ISBN: 978-3-540-87833-9

DOI 10.1007/978-3-540-87833-9

Lecture Notes in Physics ISSN: 0075-8450   e-ISSN: 1616-6361

Cover design: Integra Software Services Pvt. Ltd.

Printed on acid-free paper

9 8 7 6 5 4 3 2 1

springer.com

Preface

Data have always been *the* driving force of natural science ever since Galileo Galilei (1564–1642). In renaissance time, he introduced the comparison of theoretical predictions with experimental data as the ultimate test of any scientific theory. This change of paradigm from a qualitative to a quantitative description of nature has shaped the way we think about how science should proceed. Today, we call it the scientific method: Start with establishing a theory and test its prediction in experiments until you find a contradiction with experimental data. Then refine your theory and start all over again.

But how do we come up with a theory to start with? There are two main approaches to this: the insight of a genius and the careful inspection of data. As examples of the former, take the idea that the entropy of an ideal gas is proportional to the log of the number of microstates accessible to the atoms of this gas in phase space as proposed by Ludwig Boltzmann (1844–1906). Or the ergodic hypothesis, generally attributed to Josiah Willard Gibbs (1839–1903), that all microstates compatible with the energy of a system are equally probable. Or even more strikingly, Sir Isaac Newton's (1643–1727) idea that the force which keeps the earth orbiting the sun and the force which makes an apple fall from a tree are one and the same. Such insights are certainly motivated by, but evidently not derived from, experimental evidence. On the other hand, think of the laws of planetary motion discovered by Johannes Kepler (1571–1630). He used the extensive and accurate observations compiled by Tycho Brahe (1546–1601) to derive his theory. Other examples may be Charles Darwin (1809–1882) or Gregor Johann Mendel (1822–1884) who also relied on extensive gathering of observations to come up with their theories of evolution and inheritance.

What unites all scientists is that they seek to explain patterns in data revealing underlying principles that govern what we see in experiments. Some have great insight, others have to rely on the inspection of a large body of data to arrive at a hypothesis or theory. In the above examples, hypotheses were derived by humans. Kepler, Darwin, Mendel and many others contemplated their data often for years until they saw a common pattern. Today, computers facilitate this task. Genome databases are scanned for risk factors for certain diseases, banks let computer algorithms analyze all data available on their customers to assess their credit risk and online vendors try to detect fraud in an automated fashion to name but a few examples.

Besides the facilitation of the search for patterns connected with a specific question, more importantly, computers have enabled us to analyze data even *without* a specific question in mind. We can let computers mine data asking whether there is any pattern or structure at all. The scientist must then answer the question what brings about this structure. Naturally, this brings us to the question of what do we consider to be a pattern or regularity. In this work, we will understand everything that cannot be explained by randomness as a pattern. The structures we will be concerned with are characterized by a maximal deviation from random null models.

Specifically, we will be concerned with one particular aspect of pattern recognition and data mining: that of clustering and dimensionality reduction. With the ever increasing amount of empirical information that scientists from all disciplines are dealing with, there exists a great need for robust, scalable and easy to use clustering techniques for data abstraction, dimensionality reduction or visualization to cope with and manage this avalanche of data. The usefulness of statistical physics in solving these problems was recognized as early as in the 1950s – long before computers became as abundant as they are today [1, 2]. This monograph will show that even today, methods and in particular spin models from statistical mechanics can help in resolving – and more importantly – in understanding the related statistical inference problems.

Clustering techniques belong to the field of unsupervised learning [3]. Given a data set, the goal is to group the data points such that the points within a cluster are similar to each other and dissimilar to the rest of the data points [4–7]. The more similar the objects within a cluster are and the more different the clusters are, the better the clustering. Instead of having to deal with many individual objects, the researcher can then concentrate on the properties of only a few clusters or classes of objects with distinct features. Though intuitively clear, clustering represents a technically ill-posed problem for a number of reasons. First, it is not clear at which level of detail a cluster structure is to be defined, i.e., what is the number of clusters in a data set or whether a subset of data points shall be regarded as one cluster or be divided further. It may also be debated whether data points may belong to more than one cluster. Part of this problem is that the term "cluster" does not have a well-defined meaning. Second, it is not clear what an appropriate similarity or dissimilarity measure is supposed to be. Third and most importantly, it is difficult to differentiate whether one is not only finding what one is searching for, i.e., all clustering techniques will find some cluster structures even in random, unstructured data.

Because of these problems there exists no single clustering technique for all types of data and clustering techniques are still the subject of ongoing research. The simplest case is most likely multivariate data where each of the data points is characterized by a D-dimensional feature vector containing real valued entries. A typical example would be a set of objects characterized by a number of measurements. Then, a natural similarity measure would be the euclidean distance. As a naive approach, one can now compute the distance between all pairs of data points and successively join close data points to clusters. This is impractical for large data sets as the number of pairwise distances scales as the square of the number of data points. The method of choice then is to introduce prototypical data points as cluster centers and find the position of these cluster centers such that they represent the data set in some optimal way. To do this, only the determination of the distance of each data point from each cluster center is necessary which makes the computational effort linear in the number of data points for a given number of clusters. This approach is taken by the k-means algorithm which is probably the most widely used clustering technique despite its many shortcomings [4, 5].

Note that the introduction of prototypical data points which are representative of a cluster is only possible when an actual distance measure exists between the data points. It is not possible for instance when a matrix of pairwise similarities alone is given. This, however, is often the case.

Another problem, known as the "curse of dimensionality" [8], arises when the dimension D of the data set to be clustered increases [9]. The reason is that the data points become increasingly sparse as the dimensionality increases and the relative difference between the closest and the farthest point from an independently selected point in the data set goes to zero with increasing dimension [9, 10].

Both of these problems arise intrinsically when dealing with relational data. Here, the objects to be clustered are characterized by some sort of relation. Typically, these relations are present or known for only a small fraction of the pairs of objects and can be represented as graphs or networks. The nodes in these networks represent the objects and their relations are represented by their connections. A typical example is the set of authors of a number of scientific articles and the relation between them is whether or not they have co-authored an article together. Such data are intrinsically sparse and often the average distance (defined as the number of steps in the network) between two arbitrarily chosen nodes scales as the logarithm of the systems size, i.e., every object is close to every other object. Further, if the graph is connected, objects in different clusters will often be only the minimal distance of one step away from each other. There is no way to introduce prototypical objects as only pairwise relations are given.

While in the past multivariate data sets have dominated the applications, an increasing use and availability of data warehousing technology allows access to more and more relational data sets. Another aspect is that the first level of description for many complex systems is through the topology of their interactions, i.e., networks again. Network clustering techniques hence do not only represent exploratory data analysis tools but also are a first step in understanding complex systems [11]. Since conventional clustering techniques are inadequate for networks, a number of novel approaches have been developed in recent years [12, 13]. Despite the many efforts, a number of issues remain.

An "ideal" clustering procedure for graphs should be computationally efficient in order to deal with very large data sets, i.e., it should be able to exploit the sparsity of the data. At the same time, it should be accurate. If there is a trade-off between runtime and accuracy this should be easily mediated. It should further allow for overlapping as well as hierarchical clusters and allow to set the level of detail at which a clustering should be performed. It should have only few parameters and these should have an intuitive meaning. There should exist a precise interpretation of the clusters found, independent from the clustering technique. And most importantly, an ideal clustering procedure should provide a measure of how strong the cluster structure found deviates from that found in equivalent random data. While none of the presently available algorithms is able to combine all of these features, the present monograph is intended to provide, analyze and show the application of a clustering procedure ideal in these ways.

Chapter 1 will give a short introduction into some graph theoretical terms necessary for the discussions to follow and provide a brief overview over some important aspects of complex network study. It will illustrate the problem of structure recognition again and underline the importance of novel tests for statistical significance.

Chapter 2 then reviews a number of cluster definitions from different fields of science and surveys the current state of the art in graph clustering including discussions of the merits and shortcomings of each method.

After this discussion, a first principles approach to graph clustering in complex networks is developed in Chapters 3 and 4. In particular, the problem of structure detection is tackled via an analogy with a physical system by mapping it onto finding the ground state of an infinite range Potts spin glass Hamiltonian. The ground state energy of this spin glass is equivalent to the quality of the clustering with lower energies corresponding to better clusterings. Benchmarks for the accuracy in comparison with other methods are given. Computationally efficient update rules for optimization routines are given which work only on the links of the network and hence take advantage of the sparsity of the system. Model systems with overlapping and hierarchical cluster structures are studied. The equivalence of graph clustering and graph partitioning is derived from the fact that random graphs cluster into equally sized parts. Using known results for the expected cut size in graph partitioning for dense graphs with a Poissonian degree distribution, expectation values for the modularity of such networks are derived.

In order to extend these results to dense random networks with arbitrary degree distribution, the replica method is used in Chapter 5. The modularity of graphs with arbitrary degree distributions is calculated which allows the comparison of modularities found in real world networks and random null models. The results can also be used to improve results for the expected cut size of graph partitioning problems.

Chapter 6 is devoted to the study of modularity in sparse random networks of arbitrary degree distribution via the cavity method. This approach is complementary to the replica method and improves its results in cases of small average connectivities. Furthermore, it allows the calculation of the maximum achievable accuracy of a clustering method and provides insight into the theoretical limitations of data-driven research.

In Chapter 7, the newly developed clustering method is applied to two real world networks. The first example is an analysis of the structure of the world trade network across a number of different commodities on the level of individual countries. The second application deals with a large market network and studies the segmentation of the individual users of this market. The application shows how a network clustering process can be used to deal with large sparse data sets where conventional analyses fail.

The concluding Chapter 8 summarizes work and hints on directions for further research.

Last but not least, it is my pleasure to acknowledge a number of coworkers and colleagues for their contributions in making this work possible. I am grateful for the fruitful collaboration with Stefan Bornholdt, Peter Ahnert, Michele Leone and Douglas R. White. I have greatly enjoyed and benefited from many discussions with Konstantin Klemm, Holger Kirsten, Stefan Braunewell, Klaus Pawelzik, Albert Diaz-Guilera, Ionas Erb, Peter Stadler, Francesco Rao, Amadeo Caflisch, Geoff Rodgers, Andreas Engel, Riccardo Zecchina, Wolfgang Kinzel, David Saad and Georg Reents. Many parts of this work would not have been possible without the great expertise and support of these colleagues.

Bremen/Würzburg,   *Jörg Reichardt*
May 2008

References

1. E. T. Jaynes. Information theory and statistical mechanics. *Physical Review*, 106(4):620–630, 1957.
2. E. T. Jaynes. Information theory and statistical mechanics ii. *Physical Review*, 108(2):171–190, 1957.
3. A. Engel and C. Van den Broeck. *Statistical Mechanics of Learning*. Cambridge University Press, New York, 2001.
4. L. Kaufman and P. J. Rousseeuw. *Finding Groups in Data: An Introduction to Cluster Analysis*. Wiley-Interscience, New York, 1990.
5. B.S. Everitt, S. Landau, and M. Leese. *Cluster Analysis*. Arnold, London, 4 edition, 2001.

6. A. K. Jain, M. N. Murty, and P. J. Flynn. Data clustering: A review. *ACM Computing Surveys*, 31(3):264–323, 1999.

7. P. Arabie and L. J. Hubert. Combinatorial data analysis. *Annual Review of Psychology*, 43:169–203, 1992.

8. R. Bellman. *Adaptive Control Processes: A Guided Tour*. Princeton University Press, Princeton, 1961.

9. M. Steinbach, L. Ertöz, and V. Kumar. *New Vistas in Statistical Physics – Applications in Econo-physics, Bioinformatics, and Pattern Recognition*, chapter Challenges of clustering high dimensional data. Springer-Verlag, Berlin Heidelberg, 2003.

10. K. Beyer, J. Goldstein, R. Ramakrishnan, and U. Shaft. When is 'nearest neighbor' meaningful? In *Proceedings of 7th International Conference on Database Theory (ICDT-1999)*, pp. 217–235, Jerusalem, Israel, 1999.

11. Special Issue. Complex systems. *Science*, 284:2–212, 1999.

12. M. E. J. Newman. Detecting community structure in networks. *European Physical Journal B*, 38:321, 2004.

13. L. Danon, J. Dutch, A. Arenas, and A. Diaz-Guilera. Comparing community structure indentification. *Journal of Statistical Mechanics*, P09008, 2005.

Contents

1

Introduction to Complex Networks

Classical physics traditionally treats problems at two very different scales. On one hand, there is the microscopic scale in which the properties of and interactions between all particles involved are known and the temporal development of the system is described by a set of differential equations. Celestial mechanics is an example of this type of treatment. On the other hand, there is statistical mechanics. There also the interactions between all particles are exactly known, but the system is comprised of so many particles that the solution of the differential equations becomes not only impossible but also meaningless. The general treatment here is to either neglect all interactions or to subsume them all into an "effective field" and then deal with an effective single particle problem. Instead of exact results, only expectation values are obtained. Due to the large numbers of particles involved, these expectation values are near exact approximations. Examples of this treatment are the thermodynamics of ideal gasses or simple models of magnetic materials such as the Ising model. Both of these approaches are reductionist: the system can be described completely from the bottom-up via the properties and interactions of elementary constituents at the microscopic level.

Despite the success of the reductionist approach, a number of systems resist such description. They are too complicated to be described exactly, or insufficient information exists to be able to describe them exactly, but they are not large or simple enough that they could be reduced to effective single particle problems. Often, the constituents of the system are very heterogeneous and many different types of interactions exist. Such systems exhibit characteristic properties which are, however, not readily explained by their microscopic properties and are often called "complex" [1]. The prototypical example of a complex system is the brain. Though the workings of an individual neuron or synapse are very well understood, the mechanisms by which memory, learning, creativity or consciousness emerge from the interactions of many neurons remain largely unexplained. Complex systems also occur in other areas of biology, economies and societies and many other fields.

Reichardt, J.: *Introduction to Complex Networks*. Lect. Notes Phys. **766**, 1–11 (2009)
DOI 10.1007/978-3-540-87833-9_1   © Springer-Verlag Berlin Heidelberg 2009

A first step in understanding complex systems is trying to understand patterns and regularities of interactions in a way which might make it possible to break the systems down into possible subcomponents. To do so, it is necessary to find a way of representing complex systems.

A convenient way to represent complex systems is through graphs or networks. The interactions of the microscopical entities of the system are represented by the connections of the network. Hence, one can use the mathematical language of graph theory [2] to describe complex systems and to investigate the topological properties of the interactions defining the system. The interested reader will find three excellent and well-readable review papers which may serve as a summary and starting point into the research of complex networks [3–5]. A more comprehensive overview can be found in the following books [6–8] while an introduction for the general reader may be found in [9–11]. Before going into the discussion in detail, a number of important graph theoretical terms and relations used throughout the text shall be introduced. The reader already familiar with network analysis and basic graph theory may skip this section.

1.1 Graph Theoretical Notation

Mathematically, a network is represented as a graph $\mathcal{G}(V, E)$, i.e., an object that consists of a set of nodes or vertices V representing the objects or agents in the network and a set E of edges or links or connections representing the interactions or relations of the nodes. The cardinality of these sets, i.e, the number of nodes and edges, is generally denoted by N and M, respectively. One may assign different values w_{ij} to the links between nodes i and j in E, rendering an edge weighted or otherwise non-weighted ($w_{ij} = 1$ by convention, if one is only interested in the presence or absence of the relation). The number of connections of node i is denoted by its degree k_i. One can represent the set of edges conveniently in an $N \times N$ matrix A_{ij}, called the adjacency matrix. $A_{ij} = w_{ij}$ if an edge between node i and j is present and zero otherwise. Relations may be directed, in which case A_{ij} is non-symmetric ($A_{ij} \neq A_{ji}$) or undirected in which case A_{ij} is symmetric. Here, we are mainly concerned with networks in which self-links are absent ($A_{ii} = 0, \ \forall i \in V$). In case of a directed network, A_{ij} denotes an outgoing edge from i to j. Hence, the outgoing links of node i are found in row i, while the incoming links to i are found in column i. For undirected networks, it is clear that $\sum_{j=1}^{N} A_{ij} = k_i$. For directed networks, $\sum_{j=1} A_{ij} = k_i^{out}$ is the out-degree and equivalently $\sum_{j=1} A_{ji} = k_i^{in}$ is the in-degree of node i. It is understood that in undirected networks, the sum of degrees of all nodes in the network equals twice the number of edges $\sum_{i=1}^{N} k_i = 2M$. The distribution of the number of connections per node is called degree distribution $P(k)$ and denotes the probability that a randomly chosen node from the network has degree k. The average degree in the network is denoted by $\langle k \rangle$ and one has $N\langle k \rangle = 2M$. One can define a probability $p = 2M/N(N-1) = \langle k \rangle/(N-1)$ as the probability that an edge exists between two randomly chosen nodes from the network.

An (induced) subgraph is a subset of nodes $v \subseteq V$ with n nodes and edges $e \subseteq E$ connecting only the nodes in v. A path is a sequence of nodes, subsequent nodes in the sequence being connected by edges from E. A node i is called reachable from node j if there exists a path from j to i. A subgraph is said to be connected if every node in the subgraph is reachable from every other. The number of steps (links) in the shortest path between two nodes i and j is called the geodesic distance $d(i,j)$ between nodes i and j. A network is generally not connected, but may consist of several connected components. The largest of the shortest path distances between any pair of nodes in a connected component is called the diameter of a connected component. The analysis in this monograph shall be restricted to connected components only since it can be repeated on every single one of the connected components of a network. More details on graph theory may be found in the book by Bollobás [2].

With these notations and terms in mind, let us now turn to a brief overview of some aspects of physicists research on networks.

1.2 Random Graphs

For the study of the topology of the interactions of a complex system it is of central importance to have proper random null models of networks, i.e., models of how a graph arises from a random process. Such models are needed for comparison with real world data. When analyzing the structure of real world networks, the null hypothesis shall always be that the link structure is due to chance alone. This null hypothesis may only be rejected if the link structure found differs significantly from an expectation value obtained from a random model. Any deviation from the random null model must be explained by non-random processes.

The most important model of a random graph is due to Erdős and Rényi (ER) [12]. They consider the following two ensembles of random graphs: $\mathcal{G}(N,M)$ and $\mathcal{G}(N,p)$. The first is the ensemble of all graphs with N nodes and exactly M edges. A graph from this ensemble is created by placing the M edges randomly between the $N(N-1)/2$ possible pairs of nodes. The second ensemble is that of all graphs in which a link between two arbitrarily chosen nodes is present with probability p. The expectation value for the number of links of a graph from this ensemble is $\langle M \rangle = pN(N-1)/2$. In the limit of $N \to \infty$, the two ensembles are equivalent with $p = 2M/N(N-1)$. The typical graph from these ensembles has a Poissonian degree distribution

$$P(k) = e^{-\langle k \rangle} \frac{\langle k \rangle^k}{k!}. \tag{1.1}$$

Here, $\langle k \rangle = p(N-1) = 2M/N$ denotes the average degree in the network.

The properties of ER random graphs have been studied for considerable time and an overview of results can be found in the book by Bollobás [13]. Note that the equivalence of the two ensembles is a remarkable result. If all networks with a given number of nodes and links are taken to be equally probable, then the typical graph from this ensemble will have a Poissonian degree distribution. To draw a graph with a non-Poissonian degree distribution from this ensemble is highly improbable, unless there is a mechanism which leads to a different degree distribution. This issue will be discussed below in more detail.

Another aspect of random networks is worth mentioning: the average shortest path between any pair of nodes scales only as the logarithm of the system size. This is easily seen: Starting from a randomly chosen node, we can visit $\langle k \rangle$ neighbors with a single step. How many nodes can we explore with the second step? Coming from node i to node j via a link between them, we now have $d_j = k_j - 1$ options to proceed. Since we have k_j possible ways to arrive at node j, the average number of second step neighbors is hence $\langle d \rangle = \sum_{k=2}^{\infty}(k-1)kP(k)/(\sum_k^{\infty} kP(k)) = \langle k^2 \rangle/\langle k \rangle - 1$. Hence, after two steps we may explore $\langle k \rangle \langle d \rangle$ nodes and after m steps $\langle k \rangle \langle d \rangle^{m-1}$ nodes which means that the entire network may be explored in $m \approx \log N$ steps. This also shows that even in very large random networks, all nodes may be reached with relatively few steps. The number $d = k - 1$ of possible ways to exit from a node after entering it via one of its links is also called the "excess degree" of a node. Its distribution $q(d) = (d + 1)P(d + 1)/\langle k \rangle$ is called the "excess degree distribution" and plays a central role in the analysis of many dynamical phenomena on networks. Note that our estimate is based on the assumption that in every new step we explore $\langle d \rangle$ nodes which we have not seen before! For ER networks, though, this is a reasonable assumption. However, consider a regular lattice as a counterexample. There, the average shortest distance between any pair of nodes scales linearly with the size of the lattice.

1.3 Six Degrees of Separation

The question of short distances was one of the first addressed in the study of real world networks by Stanley Milgram [14]. It was known among sociologists that social networks are characterized by a high local clustering coefficient:

$$c_i = \frac{2m_i}{k_i(k_i - 1)}. \tag{1.2}$$

Here, m_i is the number of connections among the k_i neighbors of node i. In other words, c_i measures the probability of the neighbors of node i being connected, that is, the probability that the friends of node i are friends among each other. The average of this clustering coefficient over the set of nodes in the network is much higher in social networks than for ER random networks with the same number of nodes and links where $\langle c \rangle = p$. This would mean

that the average shortest distance between randomly chosen nodes in social
networks may not scale logarithmically with the system size. To test this,
Milgram performed the following experiment: He gave out letters in Omaha,
Nebraska, and asked the initial recipients of the letters to give the letters only
to acquaintances whom they would address by their first name and require
that those would do the same when passing the letter on. The letters were
addressed to a stock broker living in Boston and unknown to the initial recipients of the letter. Surprisingly, not only did a large number of letters arrive
at the destination, but the median of the number of steps it took was only
6. This means the path lengths in social networks may be surprisingly short
despite the high local clustering. Even more surprisingly, the agents in this
network are able to efficiently navigate messages through the entire network
even though they only know the local topology. After this discovery, Dun-
can Watts and Steve Strogatz [15] provided the first model of a network that
combines the high clustering characteristic for acquaintance networks and the
short average path lengths known from ER random graphs. At the same time,
it retains the fact that there is only a finite number of connections or friends
per node in the network. The Watts/Strogatz model came to be known as
the "small world model" for complex networks. It basically consists of a regular structure producing a high local clustering and a number of randomly
interwoven shortcuts responsible for the short average path length. It was
found analytically that a small number of shortcuts, added randomly, suffice
to change the scaling of the average shortest path length from linear with
system size to logarithmically with system size.

1.4 Scale-Free Degree Distributions

With the increasing use of the Internet as a source of information and means of
communication as well as the increasing availability of large online databases
and repositories, more and more differences between real world networks and
random graphs were discovered. Most strikingly was certainly the observation
that many real world networks have a degree distribution far from Poissonian
with heavy tails which rather follows a log-normal distribution or alternatively
a power law.

For networks with a power-law degree distribution the notion of a "scale-
free" degree distribution is often used. A scale-free degree distribution is characterized by a power law of the form

$$P(k) \propto k^{-\gamma}, \tag{1.3}$$

with some positive exponent γ. The probability of having k neighbors is in-
versely proportional to k^γ. The name "scale free" comes from the fact that
there is no characteristic value of k. While in ER graphs, the characteristic
k is the average degree $\langle k \rangle$, i.e., the average is also a typical k, there is no
typical degree in scale-free networks.

From these observations, it became clear that the assumption of equal linking probability for all pairs of nodes had to be dropped and that specific mechanisms had to be sought which could explain the link pattern of complex networks from a set of rules. Until now, many such models have been introduced which model networks to an almost arbitrary degree of detail. The starting point for this development was most likely the model by Barabási and Albert [16]. They realized that for many real world networks, two key ingredients are crucial: growth and preferential attachment, i.e., nodes that already have a large number of links are more likely to acquire new ones during the growth of the network. These two simple assumptions lead them to develop a network model which produces a scale-free degree distribution of exponent $\gamma = 3$. Consequently, this model was used as a first attempt to explain the link distribution of web pages.

In order to model an ensemble of random graphs with a given degree distribution without resorting to some growth model of how the graph is knit the "configuration model" can be used. It is generally attributed to Molloy and Reed [17], who devised an algorithm for constructing actual networks, but it was first introduced by Bender and Canfield [18]. The configuration model assumes a given degree distribution $P(k)$. Every node i is assigned a number of stubs k_i according to its degree drawn from $P(k)$ and then the stubs are connected randomly. For this model, the probability that two randomly chosen nodes are connected by an edge can be well approximated by $p_{ij} = k_i k_j / 2M$ as long as the degrees of the nodes are smaller than $\sqrt{2M}$. The probability to find a link between two nodes is hence proportional to the product of the degrees of these two nodes. The configuration model and the ER model make fundamentally different assumptions on the nature of the objects represented by the nodes. In the ER model, fluctuations in the number of connections of a node arise entirely due to chance. In the configuration model, they represent a quality of the node which may be interpreted as some sort of "activity" of the object represented by the node.

1.5 Correlations in Networks

Thus far, only models in which all nodes were equivalent have been introduced. In many networks, however, nodes of different types coexist and the probability of linking between them may depend on the types of nodes. A typical example may be the age of the nodes in a social network. Agents of the same age generally have a higher tendency to interact than agents of different ages. Let us assume the type of each node is already known. One can then ask whether the assumption holds, that links between nodes in the same class are indeed more frequent than links between nodes in different classes. Newman [19] defines the following quantities: e_{rs} as the fraction of edges that fall between nodes in class r and s. Further, he defines $\sum_r e_{rs} = a_s$ as the fraction of edges that are connected to at least one node in class s. Note that

e_{rs} can also be interpreted as the probability that a randomly chosen edge lies between nodes of class r and s and that a_s can be interpreted as the probability that a randomly chosen edge has at least one end in class s. Hence, a_s^2 is the expected fraction of edges connecting two nodes of class s. Comparing this expectation value with the true value e_{ss} for all groups s leads to the definition of the "assortativity coefficient" r_A:

$$r_A = \frac{\sum_s \left(e_{ss} - a_s^2\right)}{1 - \sum_s a_s^2}. \tag{1.4}$$

This assortativity coefficient r_A is one, if all links fall exclusively between nodes of the same type. Then the network is perfectly "assortative", but the different classes of nodes remain disconnected. It is zero if $e_{ss} = a_s^2$ for all classes s, i.e., no preference in linkage for either the same or a different class is present. It takes negative values, if edges lie preferably between nodes of different classes, in which case the network is called "disassortative". The denominator corresponds to a perfectly assortative network. Hence, r_A can be interpreted as the percentage to which the network is perfectly assortative.

For the classes of the nodes, any measurable quantity may be used [20]. Especially interesting are investigations into assortative mixing by degree, i.e., do nodes predominantly connect to other nodes of similar degree (assortative, $r_A > 0$) or is the opposite the case (disassortative, $r_A < 0$). It was found that many social networks are assortative, while technological or biological networks are generally disassortative [20]. Note that r_A may also be generalized to the case where the class index s takes continuous values [20]. It should be stressed that such correlation structures do not affect the degree distribution.

1.6 Dynamics on Networks

Apart from these topological models mainly concerned with link structure, a large number of researchers are concerned with dynamical processes taking place on networks and the influence the network structure has on them. Among the most widely studied processes is epidemic spreading and one of the most salient results is certainly that by Cohen [21, 22], which shows that for scale-free topologies with exponents larger than two and low clustering, the epidemic threshold (the infectiousness a pathogen needs to infect a significant portion of the network) drops to zero. The reason for this is, in principle, the fact that for scale-free degree distributions with exponents between 2 and 3 the average number of second neighbors $\langle d \rangle$ may diverge. Liljeros showed that networks of sexual contacts do have indeed such a topology [23]. At the same time, these results brought about suggestions for new vaccination techniques such as the vaccination of acquaintances of randomly selected people which allows us to vaccinate people with higher numbers of connections with higher efficiency [24]. Consequently, a number of researchers are also studying the interplay between topology of the network and dynamic processes on networks in models that allow dynamic rewiring of connections in accordance with, for instance, games being played on the network to gain insights into the origin of cooperation [25].

All of this research has shown the profound effect of the topology of the connections underlying a dynamical process and hence underlines the importance of thoroughly studying the topology of complex networks.

1.7 Patterns of Link Structure

The above discussion has shown the importance of investigating the link structure in real world networks. One can view this problem as a kind of pattern detection. Patterns are generally viewed as expressions of some kind of regularity. What such a regularity may be, however, remains often a vague concept. It might be sensible to define everything as regular which is *not* random.

The structure this monograph is concerned with is a particular type of non-random structure in complex networks which is closely related to the aforementioned correlations. The section about correlations has shown that if the different types of nodes in a network are known, the link structure of the network may show a particular signature. In the majority of cases, however, the presence of different types of nodes is only hypothesized and the type of each node is unknown. The purpose of this work is to develop methods to detect the presence of different types of nodes in networks and to find the putative type of each node. A number of possible applications from various fields shall motivate the problem again.

Suppose we are given a communication network of an enterprise. Nodes are employees and links represent communication, e.g., via e-mail, between them. We may then search for "communities of practice" – employees who are particularly well connected among each other, i.e., with highly enriched in-group communication. It is then possible to compare these communities of practice to the organizational structure of the enterprise and possibly use the results in the assembly of teams for future projects. A study in this direction has been performed by Tyler et al. [26].

Novel experimental techniques from biology allow the automatic extraction of all proteins produced by an organism. Proteins are the central building blocks of biological function, but generally, proteins do not function alone but bind to one another to form complexes which in turn are capable of performing a particular function, such as initiating the transcription of a particular piece of DNA. It is now possible to study the pairwise binding interactions of a large number of proteins in an automated way [27]. The result of such a study is a protein interaction network in which the links represent pairwise interactions between proteins. Protein function should be mirrored in such a network. For instance, proteins forming part of a complex should now be detectable as densely interlinked groups of nodes in such a network [28]. An analysis of the structure of a protein interaction or other biological network created by automated experiments hence presents a first step in planning future experiments [29].

The collection of scientific articles represent a strongly fragmented repository of scientific knowledge [30, 31]. Online databases make it possible to study these in an automated way, e.g., in form of co-authorship networks or citation networks. In the former, nodes are researchers while links represent co-authorship of one or more articles. Analysis of the structure of this network may give valuable information about the cooperation between various scientists and aid in the evaluation of funding policy or influence future funding decisions. In the latter, nodes in the network are scientific articles and links denote the citation of one from the other. Analysis of the structure of this network may yield insight into the different research areas of a particular field of science.

With these examples in mind it becomes clear that the detection of structural patterns is not only important in the description of complex systems which are represented through networks, but can be viewed as an elementary technique for the exploratory analysis of any kind of relational data set. The above examples have also illustrated that such exploratory analysis is often the starting point of further work. It is therefore important to assess the statistical validity of the findings and avoid the "deception of randomness" [32], i.e., to ensure that the findings of an algorithm are statistically significant and not the mere result of the search process. To illustrate this, let us consider the following problem: Given is an ER network with average degree $\langle k \rangle = 5$. Also given is that the network consists of two types of nodes A and B with 50 nodes of each type and only 42 links running between nodes of type A and B and the remainder of the 250 links within groups of type A and B, respectively. If nodes were connected independently of their type, the total number of links between type A and B nodes is Poisson distributed with a mean of $\langle k \rangle N/4 = 125$ and a standard deviation of $\sigma = 11$. Hence, finding only 42 links between A and B is statistically highly significant with a p-value of $p = 2.8 \times 10^{-18}$. Now assume that the type of each node was not given and we had searched for an assignment into two equally sized groups A and B with a minimum number of links between the groups. At which number of links between groups would such a finding become significant? The search space for this task consists of $\binom{N}{N/2}$ possible assignments of the nodes into two equally sized groups. Applying a Bonferroni correction [33] for this number of different "experiments" would lead to a situation where less than 22 connections between nodes of type A and B would be significant at the 5% level. With this correction, recovering the initial configuration with 42 links between nodes of different types could not be called significant. As will be shown in the course of this work, 42 is the typical number of links running between the different parts when partitioning an ER random network with 100 nodes and $\langle k \rangle = 5$ into two equal sized groups. In other words, every configuration with less than 42 links between groups indicates a significant deviation from pure random linking and hence structure in the data. Thus, statistical significance starts much earlier than the limit given by the Bonferroni correction. The Bonferroni correction fails here because it assumes independent experiments. The different assignments into groups produced by a search process, however, are not independent.

These considerations should exemplify that standard statistical tests are problematic when the assignment of types to nodes results from a search process and novel methods for the assessment of statistical significance are needed. Large parts of this work are therefore devoted to the study of what kind of structure can be found in random networks such that everything beyond these expectation values for random networks can be taken as indicating true structure in the data.

References

1. Special Issue. Complex systems. *Science*, 284:2–212, 1999.
2. B. Bollobás. *Modern Graph Theory*. Springer-Verlag Berlin Heidelberg, 1998.
3. R. Albert and A. -L. Barabàsi. Statistical mechanics of complex networks. *Reviews of Modern Physics*, 74:47–97, 2002.
4. M. E. J. Newman. The structure and function of complex networks. *SIAM Review*, 45(2):167–256, 2003.
5. S. N. Dorogovtsev and J. F. F. Mendes. The shortest path to complex networks. In N. Johnson, J. Efstathiou, and F. Reed-Tsochas, editors, *Complex Systems and Inter-disciplinary Science*. World Scientific, Singapore, 2005.
6. S. Bornholdt and H. G. Schuster, editors. *Handbook of Graphs and Networks*. Wiley-Vch, Weinheim, 2003.
7. E. Ben-Naim, H. Frauenfelder, and Z. Toroczkai. *Complex Networks, Lect. Notes Phys. 650*. Springer-Verlag, Berlin Heidelberg, 2004.
8. J. Mendes. *Science of Complex Networks. From Biology to the Internet and WWW*. Springer-Verlag, Berlin Heidelberg, 2005.
9. M. Buchanan. *Nexus: Small Worlds and the Groundbreaking Science of Networks*. W.W. Norton & Company, New York, 2003.
10. D. Watts. *Six Degrees: The Science of a Connected Age*. W.W. Norton & Company, New York, 2004.
11. A. -L. Barabási. Linked. Plume Books, New York, 2003.
12. P. Erdős and A. Rényi. On the evolution of random graphs. *Publications of the Mathematical Institute of the Hungarian Academy of Sciences*, 5:17–61, 1960.
13. B. Bollobas. *Random Graphs*. Cambridge University Press, New York, 2nd edition, 2001.
14. S. Milgram. The small world problem. *Psychology Today*, 2:60–67, 1967.
15. D. J. Watts and S. H. Strogatz. Collective dynamics in small world networks. *Nature*, 393:440–442, 1998.
16. A. -L. Barabási and R. Albert. Emergence of scaling in random networks. *Science*, 286:509–512, 1999.
17. M. Molloy and B. Reed. A critical point for random graphs with given degree sequence. *Random Structures and Algorithms*, 6:161–179, 1995.

18. E. A. Bender and E. R. Canfield. The asymptotic number of labeled graphs with given degree distribution. *Journal of Combinatorial Theory A*, 24:296, 1978.

19. M. E. J. Newman. Assortative mixing in networks. *Physical Review Letters*, 89:208701, 2002.

20. M. E. J. Newman. Mixing patterns in networks. *Physical Review E*, 67:026126, 2003.

21. R. Cohen, K. Erez, and S. Havlin. Resilience of the internet to random breakdowns. *Physical Review Letters*, 85:4626–4628, 2000.

22. R. Pastor-Satorras and A. Vespignani. Epidemic spreading in scale-free networks. *Physical Review Letters*, 86:3200–3203, 2001.

23. F. Liljeros, C. R. Edling, L. A. N. Amaral, H. E. Stanley, and X. Aberg. The web of human sexual contacts. *Nature*, 411:907–908, 2001.

24. R. Pastor-Satorras and A. Vespignani. Immunization of complex networks. *Physical Review E*, 036104, 2002.

25. G. Szabo and G. Fath. Evolutionary games on graphs. *pre-print cond-mat/0607377*, 2006.

26. J. R. Tyler, D. M. Wilkinson, and B. A. Huberman. Email as spectroscopy: automated discovery of community structure within organisations. In *International Conference on Communities and Technologies*, Amsterdam, The Netherlands, 2003.

27. P. Uetz, L. Giot, G. Cagney, T. A. Mansfield, R. S. Judson, J. R. Knight, D. Lockshon, V. Narayan, M. Srinivasan, P. Pochardt, A. Qureshi-Emili, Y. Li, B. Godwin, D. Conver, T. Kalbfleisch, G. Vijayadamodar, M. Yang, M. Johnston, S. Fields, and J. M. Rothberg. A comprehensive analysis of protein–protein interactions in saccharomyces cerevisiae. *Nature*, 403:623–627, 2000.

28. V. Spirin and L. A. Mirny. Protein complexes and functional modules in molecular networks. *Proceedings of the National Academy of Sciences of the United States of America*, 100(21):12123, 2003.

29. R. Guimera and L. A. N. Amaral. Functional cartography of complex metabolic networks. *Nature*, 433:895–900, 2005.

30. various authors. Special issue: Mapping knowledge domains. *Proceedings of the National Academy of Sciences of the United States of America*, 101, 2004.

31. M. E. J. Newman. Scientific collaboration networks: I. network construction and fundamental results. *Physical Review E*, 64:016131, 2001.

32. A. Engel and C. Van den Broeck. *Statistical Mechanics of Learning*. Cambridge University Press, New York, 2001.

33. C.E. Bonferroni. Il calcolo delle assicuranzioni su gruppi die teste. *Studi in Onore del Professore Salvatore Ortu Carboni*, pp. 13–60, 1935.

2

Standard Approaches to Network Structure: Block Modeling

2.1 Positions, Roles and Equivalences

By investigating data from a wide range of sources encompassing the life
sciences, ecology, information and social sciences as well as economics, researchers have shown that an intimate relation between the topology of a
network and the function of the nodes in that network indeed exists [1–9].
A central idea is that nodes with a similar pattern of connectivity will perform a similar function. Understanding the topology of a network will be a
first step in understanding the function of individual nodes and eventually the
dynamics of any network.

As before, we can base our analysis on the work done in the social sciences.
In the context of social networks, the idea that the pattern of connectivity is
related to the function of an agent in the network is known as playing a "role"
or assuming a "position" [10,11]. Here, we will endorse this idea.

The nodes in a network may be grouped into equivalence classes according
to the role they play. Two basic concepts have been developed to formalize the
assignments of roles individuals play in social networks: structural and regular
equivalence. Both are illustrated in Fig. 2.1. Two nodes are called structurally
equivalent if they have the exact same neighbors [12]. This means that in the
adjacency matrix of the network, the rows and columns of the corresponding
nodes are exactly equal. The idea behind this type of equivalence is that two
nodes which have the exact same interaction partners can only perform the
exact same function in the network. In Fig. 2.1, only nodes A and B are
structurally equivalent while all other nodes are structurally equivalent only
to themselves.

To relax this very strict criterion, regular equivalence was introduced [13,
14]. Two nodes are regularly equivalent if they are connected in the same
way to equivalent others. Clearly, all nodes which are structurally equivalent
must also be regularly equivalent, but not vice versa. The seemingly circular
definition of regular equivalence is most easily understood in the following way:
Every class of regularly equivalent nodes is represented by a single node in an

Reichardt, J.: *Standard Approaches to Network Structure: Block Modeling.* Lect. Notes
Phys. **766**, 13–30 (2009)
DOI 10.1007/978-3-540-87833-9_2   © Springer-Verlag Berlin Heidelberg 2009

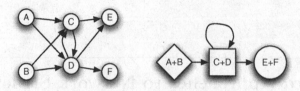

Fig. 2.1. Example network illustrating structural and regular equivalence. Nodes A and B have the same neighbors and are thus structurally equivalent and regularly equivalent. Nodes C though F form four different classes of structural equivalence but can be grouped into only two classes of regular equivalence as shown in the image graph or role model on the right.

"image graph". The nodes in the image graph are connected (disconnected), if connections between nodes in the respective classes exist (are absent) in the original network. In Fig. 2.1, nodes A and B, C and D as well as E and F form three classes of regular equivalence. If the network in Fig. 2.1 represents the trade interactions on a market, we may interpret these three classes as producers, retailers and consumers, respectively. Producers sell to retailers, while retailers may sell to other retailers and consumers, which in turn only buy from retailers. The image graph (also "block" or "role model") hence gives a bird's-eye view of the network by concentrating on the roles, i.e., the functions, only. Note that no two nodes in the image graph may be structurally equivalent, otherwise the image graph is redundant.

2.2 Block Modeling

Let us consider the larger example from Fig. 2.2. The network consists of 18 nodes in 4 designed roles. Nodes of type A connect to other nodes of type A and to nodes of type B. Those of type B connect to nodes of type A and C,

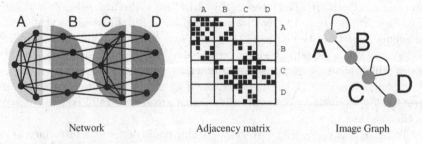

| Network | Adjacency matrix | Image Graph |

Fig. 2.2. An example network and two possible block models. The nodes in this network can be grouped into four classes of regular equivalence (A, B, C and D). Ordering the rows and columns of the adjacency matrix according to the four classes of regular equivalence makes a block structure apparent (there are 16 blocks from the 4 classes), which is efficiently represented by an image graph.

acting as a kind of intermediary. Nodes of type C have connections to nodes of type B, others of type C and of type D. Finally, nodes of type D form a kind of periphery to nodes of type C. An ordering of rows and columns according to the types of nodes makes a block structure in the adjacency matrix apparent. Hence the name "block model". The image graph effectively represents the 4 roles present in the original network and the 16 blocks in the adjacency matrix. Every edge present in the network is represented by an edge in the image graph and all edges absent in the image graph are also absent in the original network.

Regular equivalence, though a looser concept than structural equivalence, is still very strict as it requires the nodes to play their roles *exactly*, i.e., each node must have at least one of the connections required and may not have any connection forbidden by the role model. In Fig. 2.2, a link between two nodes of type A may be removed without changing the image graph, but an additional link from a node of type A to a node of type C would change the role model completely. Clearly, this is unsatisfactory in situations where the data are noisy or only approximate role models are desired for a very large data set.

One way to relax the strict condition of regular equivalence is to allow for "generalized block modeling" introduced by Doreian, Batagelj and Ferligoj [15]. Figure 2.3 summarizes the nine types of blocks introduced by Doreian et al. They encompass the different possibilities that arise in directed networks already:

- Complete blocks:
  Every node of type X has an outgoing link to every node of type Y and each node of type Y has an incoming link from every node of type X. If we have only complete and null blocks, we have found a classification into structural equivalence classes.
- Row dominant:
  One node of type X has an outgoing link to every node of type Y.
- Column dominant:
  One node of type Y has an incoming link from every node of type X.
- Regular:
  Each node of type X has at least one outgoing link to a node of type Y *and* each node of type Y has at least one incoming link from a node of type X. If all blocks in the block model are regular, complete or null, we have found a classification into regular equivalence classes.
- Row regular:
  Each node of type X has at least one outgoing link to a node of type Y.
- Column regular:
  Each node of type Y has at least one incoming link from a node of type X
- Null:
  There are no links connecting nodes of type X and Y.

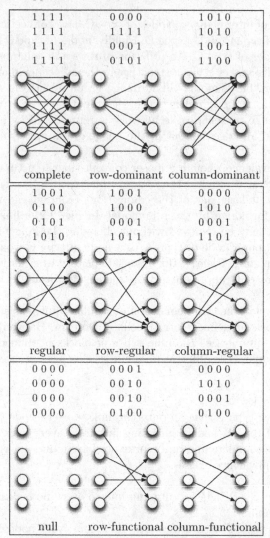

Fig. 2.3. Example adjacency matrices and corresponding connection patterns according to the nine types of blocks for generalized block modeling after [15].

- Row functional:
  Each node of type X has exactly one outgoing link to a node of type Y.
- Column functional:
  Each node of type Y has exactly one incoming link from a node of type X.

The introduction of row- and column-dominant, row- and column-regular and row- and column- functional blocks is what relaxes the condition of regular equivalence and allows for a wider spectrum of possible topologies to be summarized as a block model. For undirected networks, the need to differentiate between row- and column-type blocks vanishes and one is left with an extension of regular blocks to dominant and functional blocks.

Even with these extensions, two major problems remain: One is the fault tolerance, i.e., a bock model should be robust to both false positive as well as false negative links in the network. The second is that for generalized block modeling as introduced by Doreian et al., it is not clear how to find the best block model for a network. The general approach is to hypothesize a block structure and then try to find a good assignment of nodes into the different classes as to fit the proposed block model. To overcome these problems, we will adopt and extend the density-based approach as already suggested in [11]. It may be possible to assign nodes into groups such that blocks of high and low density of links appear in the adjacency matrix of the network. Figure 2.4 shows a number of examples of this together with the corresponding image graphs indicating between which types of nodes a high density of links exists. The adjacency matrices are larger now than in Fig. 2.2 and make the blocks resulting from the different link densities more salient. Before this, however, we will focus on one particular type of block structure: so-called cohesive blocks or modular or community structures. Modular or community structures are

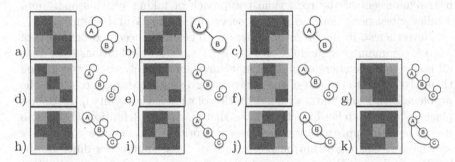

Fig. 2.4. Example adjacency matrices for networks with 60 nodes in 2 or 3 roles and corresponding image graphs. Rows and columns in matrices are ordered such that nodes of the same type (in the same role) are adjacent. Hence, blocks appear in the adjacency matrix due to the similar pattern of connectivity among nodes of the same type. The types are represented by a single node in the image graph. Background shading of matrices reflects the link density in blocks. We show only those three-role models which are not isomorphic and which cannot be reduced to a block model of two roles only. The two-role models can be understood as (**a**) modular structure, where nodes connect primarily to nodes of the same role; (**b**) bi-partition, with connections primarily between nodes of different type; and (**c**) a core–periphery structure with nodes of type A (the core) connecting preferentially among themselves and to nodes of type B (the periphery). The three-role models can be seen as combinations of these three basic structures plus the possibility of having intermediates.

characterized by the fact that the nodes in a network may be assigned into groups which are densely connected internally but only sparsely to the rest of the network. In Fig. 2.4 the networks (a) and (d) are examples of this. We may also call such modular structures "diagonal block models".

2.3 Cohesive Subgroups or Communities as Block Models

The abundance of diagonal block models or modular structures makes modularity a concept so important that it is often studied outside the general framework of block modeling. One explanation may be that in social networks it may even be the dominant blocking structure. The reason may be that homophily [16], i.e., the tendency to form links with agents similar to oneself, is a dominant mechanism in the genesis of social networks. Recall, however, that the concept of functional roles in networks is much wider than mere cohesiveness as it specifically focuses on the *inter-dependencies* between groups of nodes. Modularity or community structure, emphasizing the absence of dependencies between groups of nodes is only one special case. It may also be that the concept of modularity appeals particularly to physicists because it is reminiscent of the reductionist approach of taking systems apart into smaller subsystems that has been so successful in the natural sciences.

Nevertheless, in the literature, there is no generally accepted definition of what a community or module actually is. A variety of definitions exist that all imply that members of a community are more densely connected among themselves than to the rest of the network. Two approaches exist to tackle the problem. Either, one starts with a definition of what a community is in the first place and then searches for sets of nodes that match this definition. Or one can use a heuristic approach by designing an algorithm and define a community as whatever this algorithm outputs. Both of these approaches differ in one fundamental way: When starting from a definition of community, it often occurs that some nodes in the network will not be placed into any community. The algorithmic approaches on the other hand will generally partition the set of vertices such that all nodes are found in some community. Whether all nodes need to be assigned into a community needs to be decided by the researcher and may determine which definitions and methods are useful in the analysis of actual data. With these considerations in mind we shall briefly review the approaches taken in the literature.

2.3.1 Sociological Definitions

The study of community structure has a long tradition in the field of sociology and it comes as no surprise that the example that sparked the interest of physicists in the field was a sociological one [17, 18]. Alternatively to community, the term *cohesive subgroup* is often used to subsume a number of definitions that emphasize different aspects of the problem. These can be grouped into definitions based on reachability, nodal degree or the comparison of within to outside links [11].

Cliques are complete subgraphs, such that every member is connected to every other member in the clique. An *n-clique* is a maximal subgraph, such that the geodesic distance $d(i, j)$ between any two members i, j is smaller or equal to n. Naturally, cliques are 1-cliques. Note that the shortest path may also run through nodes not part of the n-clique, such that the diameter of an n-clique may be larger than n. An *n-clan* denotes an n-clique with diameter less or equal to n. Naturally, all n-clans are also n-cliques. Alternatively, an *n-club* is a maximal subgraph of diameter n.

These definitions are problematic in several ways. Cliques can never get larger than the smallest degree among the member nodes which limits these communities to be generally very small in large networks with limited degrees. The other definitions relying on distances are problematic if the network possesses the small world property. The overlap of such communities will generally be as large as a typical group.

Another group of definitions is based on the degree of the members of a community. A *k-plex* is a maximal subgraph of n nodes, such that each member has at least $n - k$ connections to other nodes in the k-plex. This definition is less strict than that of a clique as it allows some links to be missing. At the same time, a k-plex only contains nodes with minimum degree $d \geq (n - k)$. A *k-core* denotes a maximal subgraph, such that each node has at least k connections to other members of the k-core.

Here again, the size of k-plexes is limited by the degrees of the nodes. K-cores are problematic also because they disregard all nodes with degree smaller than k even if they have all their connections to nodes within this core.

While the two former groups of definitions are based primarily on internal connections, a number of definitions of cohesive subgroups exist which compare intra- and inter-group connections. One example are *LS sets*. A set of n nodes is an LS set, if each of its proper subsets has more ties to its complement than to the rest of the network.

The problem with this definition may be studied with an example. Assume a clique of 10 nodes in a large network. Each of the members of this clique has only 1 link to the rest of the network. This is not an LS set, because 9 of the 10 nodes taken together have 9 links to the complement in the set and also 9 links to the rest of the network. This is indeed a paradoxical situation, as every node has 9 out of 10 links to other members of the same set of nodes.

It should be noted that while nodes may be part of several n-cliques, n-clubs or n-clans, i.e., these sets may overlap, LS sets are either disjoint or one contains the other and they hence induce a hierarchy of communities in the graph [11].

Yet an alternative definition of a cohesive subgroup is the following. If the edge connectivity $\lambda(i, j)$ of two nodes i and j is defined as the minimum number of links that must be removed from the graph to leave nodes i and j disconnected, then a λ *set* denotes a maximal subgraph, such that $\lambda(i,j) > \lambda(i,k)$ for all nodes i,j which are members of the λ set and all nodes k which are not members of it.

For the lambda sets, again one has the problem that nodes of low degree, though having all of their connections with the community, may not belong to it.

2.3.2 Definitions from Physicists

The diversity of definitions from sociology already indicates the conceptual difficulties involved and demonstrates that the question of what a community is may not have a simple answer. To make things worse, a number of alternative definitions have been and continue to be contributed by physicists as well [19, 20].

Radicchi et al. [21] have introduced the notion of community *in a strong sense* and *in a weak sense*. For a subgraph V of $\mathcal{G}$ to be a community in the strong sense, they require

$$k_i^{in} > k_i^{out} \quad \forall \; i \in V, \tag{2.1}$$

i.e., the number of internal connections k_i^{in} to other members of V shall be larger than the number of external connections k_i^{out} to the rest of the network. Note that $k_i^{in} + k_i^{out} = k_i$, the degree of node i. Relaxing this condition, for a subgraph V to be a community in a weak sense they require

$$\sum_{i \in V} k_i^{in} > \sum_{i \in V} k_i^{out}. \tag{2.2}$$

A paradoxical issue arising from both of these definitions is that communities in the strong or weak sense can be formed of disconnected subgraphs as long as these subgraphs also obey the definition. It should be noted, however, that this definition was initially proposed as a stop criterion for hierarchical agglomerative or divisive clustering algorithms.

Palla *et al.* [8, 22] have given an alternative definition based on reachability, though defined through a clique percolation process and not via paths in the network. Two k-*cliques* are adjacent if they share a $(k-1)$-clique, i.e., they differ by only one node. Note that the term k-cliques here denotes complete subgraphs with k nodes. As a community or k-clique percolation cluster, they define the set of nodes connected by $(k-1)$-cliques. An example will clarify these issues. Two vertices connected by an edge form a 2-clique. Two triangles (3-cliques) are adjacent if they share an edge, i.e., a 2-clique. This definition allows nodes to be part of more than one community and hence allows for overlap among communities much like the other definitions based on reachability.

Other approaches given by physicists and computer scientists are algorithmically motivated. The next section will discuss this treatment of the problem.

2.4 Algorithms for Community Detection

One may ask how it shall be possible to design a community detection algorithm without a precise definition of community. The answer is that for many networks the community structure is known from other sources and the reasoning is that any algorithm, which is good at discovering known community structures, will be good at finding unknown ones as well. A number of real world data sets have become almost standard for this purpose and will be discussed in the following chapters and later sections.

In addition to real world networks with known community structure it has become customary to compare the performance of community detection algorithms on computer-generated test networks with known communities. The standard example is the following: Given is a graph with 128 nodes, divided into 4 communities of 32 nodes each. The degree distribution is chosen to be Poissonian with an average of $\langle k \rangle = 16$. The links of every node are divided into those that connect to other members of the same community and those connecting to the rest of the network, such that

$$\langle k \rangle = \langle k_{in} \rangle + \langle k_{out} \rangle. \tag{2.3}$$

Otherwise, the network is completely random. For fixed $\langle k \rangle$, recovering the built-in community structure becomes more difficult as $\langle k_{out} \rangle$ increases at the expense of $\langle k_{in} \rangle$. It has become customary to study the performance of an algorithm as a function of $\langle k_{in} \rangle$.

2.4.1 Comparing a Quality Function

Instead of comparing the output of an algorithm for networks with known community structure one may compare the results of different algorithms across a quality function for the assignment of nodes into communities. Newman and Girvan [23] have proposed the following measure of the "modularity" of a community structure with q groups:

$$Q = \sum_{s=1}^{q} e_{ss} - a_s^2, \text{ with } a_s = \sum_{s=1}^{q} e_{rs}. \tag{2.4}$$

Here, e_{rs} is the fraction of all edges that connect nodes in groups r and s and hence e_{ss} is the fraction of edges connecting the nodes of group s internally. From this, one finds that a_s represents the fraction of all edges having at least one end in group s and a_s^2 is to be interpreted as the expected fraction of links falling between nodes of group s given a random distribution of links. Note the similarity of this measure with the assortativity coefficient defined earlier. It is clear that $-1 < Q < 1$.

This modularity measure will play a central role in the following chapters and it is of course a natural idea to optimize the assignment of nodes in communities directly by maximizing the modularity of the resulting partition.

2.4.2 Hierarchical Algorithms

A large number of heuristic algorithmic approaches to community detection have been proposed by computer scientists. The developments follow generally along the lines of the algorithms developed for multivariate data [24–26]. Typically, the problem is approached by a recursive min-cut technique that partitions a connected graph into two parts minimizing the number of edges to cut [27,28]. These treatments, however, suffer greatly from being very skewed as the min-cut is usually found by cutting off only a very small subgraph [29]. A number of penalty functions have been suggested to overcome this problem and balance the size of subgraphs resulting from a cut. Among these are ratio cuts [29,30], normalized cuts [31] or min–max cuts [32].

The clustering algorithm devised by Girvan and Newman (GN) [17] was the first to introduce the problem of community detection to physics researchers in the field of complex networks. As is often the case, the impact the paper created was not merely for the algorithm but because of the well-chosen illustrative example of its application. GN's algorithm is based on "edge betweenness" – a concept again borrowed from sociology. Given all geodesic paths between all pairs of nodes in the network, the betweenness of an edge is the number of such paths that run across it. It is intuitive that betweenness is a measure of centrality and hence introduces a measure of distance to the graph. The GN algorithm calculates the edge betweenness for all edges in the graph and then removes the edge with the highest betweenness. Then, the betweenness values for all edges are recalculated. This process is repeated until the network is split into two disconnected components and the procedure starts over again on each of the two components until only single nodes remain. The algorithm falls into the class of recursive partitioning algorithms and its output is generally depicted as a dendrogram illustrating the progression of splitting the network.

Figure 2.5 illustrates the algorithm with the example chosen by GN [17]. The network shown displays the friendships among the members of a karate club at a US university compiled by the anthropologist Zachary [18] over a period of 2 years. Over the course of the observation an internal dispute between the manager (node 34) and the instructor of the club (node 1) led to the split up of the club. Roughly half of the members joined the instructor in the formation of a new club and the other half of the members stayed with the manager hiring a new instructor. It turns out that the first split induced by the GN algorithm corresponds almost exactly to the observed split among the members of the club. This led to the conclusion that the split could be "predicted" from the topology of the network and that the GN algorithm is able to make such predictions. As far as the definition of community is concerned, the algorithm induces a hierarchy of communities as at any level of progress of the algorithm a set of connected nodes is to be understood as a community.

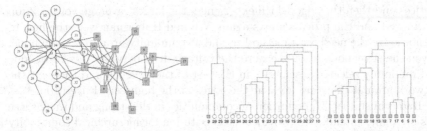

Fig. 2.5. *Left:* The karate club network due to Zachary [18]. *The circles* denote members who sided with the manager, while the *squares denote* members siding with the instructor in the split observed by Zachary. *Right:* The dendrogram output by the GN algorithm. Note that the first split of the algorithm corresponds almost exactly the split observed. Both figures from [17].

The main problem of the GN algorithm is its high demand of computational resources running in $\mathcal{O}(N^3)$ steps for networks with N nodes. Also, it is not clear at which level in the dendrogram a cut is best. The algorithm is completely deterministic and therefore error prone in case of noisy data and possible alternative community structures cannot be found.

A solution to the latter problem was suggested by Tyler *et al.* [7, 33]. Instead of calculating the edge betweenness from all shortest paths between all pairs of nodes, they merely sample the edge betweenness by calculating the edge betweenness between randomly chosen pairs of nodes. This reduces the computational complexity drastically, since instead of calculating the shortest paths between $\mathcal{O}(N^2)$ pairs of vertices, only a fraction of them needs to be sampled. Repeating the entire process, the algorithm then aggregates nodes into communities which are repeatedly in the same connected component in very late stages of the partitioning. This modification is intended to address the problem that the GN algorithm is deterministic, i.e., it is capable of producing only one output given a data set and an estimation of the stability of the community assignment with respect to the removal or addition of single links cannot be easily evaluated.

Newman has also introduced a measure of edge betweenness centrality based on random walks [34], i.e., the edge betweenness is interpreted as the number of times it is traversed by a random walker. This measure can be used for community detection in the same way as the shortest path betweenness.

In a similar and elegant way, Wu and Huberman have proposed a method to calculate the betweenness through an analogy with a resistor network [35]. The network is viewed as a resistor network, the edges being the resistors, and the betweenness of an edge is estimated from the voltage drop across the resistor when a voltage is applied between two randomly chosen connector nodes. Of course, those resistors which have few resistors in parallel will show the largest voltage drop corresponding to the largest betweenness. The voltage drops are sampled for a number of randomly chosen pairs of connector vertices and then the edge which experiences the largest average voltage drop is removed and the process starts again. Wu and Huberman were the first to acknowledge the need for a method to find a community around a given node. Given the start node, they measure the voltage drop between this start node and randomly selected test nodes in the rest of the network and then cut the network around the start node at the edges of the highest voltage drop.

Radicchi et al.'s [21] definition of communities in the strong and weak sense was originally intended to provide a stop criterion for hierarchical community detection algorithms. As an example, the GN algorithm should be stopped when the next split would result in groups of nodes that do not comply to the definitions given by Radicchi et al. In order to speed up the calculation of betweenness, Radicchi et al. also present an approximation method. From the observation that edges with high betweenness generally have few alternative paths, they define the edge clustering coefficient as

$$c_{ij} = \frac{z_{ij} + 1}{\min(k_i - 1, k_j - 1)}. \tag{2.5}$$

Here z_{ij} denotes the number of triangles above the edge connecting nodes i and j. Edges for which either k_i or k_j is zero are excluded from consideration. Their algorithm then consists in successively removing the edges with lowest edge clustering corresponding to those with highest betweenness. Note that this algorithm strongly depends on the existence of triangles. It may be extended to other loops, but keep in mind that the small world property of many networks makes such extension computationally costly.

Another way of determining when to stop a recursive partitioning algorithm is to assess the network modularity Q at every split and accept a split only when the split results in an increase of the modularity Q, since for the whole network as one community $Q = 0$, there always exists one split which increases Q. Equivalently, one can assign a different community index to every node and then successively join those pairs of nodes or groups of nodes which lead to the largest increase in Q. This is the idea behind the algorithm presented by Clauset et al. [36]. While the other algorithms presented so far are hierarchical divisive algorithms, this one starts from the bottom-up in an agglomerative manner.

A hierarchical approach based on a dynamical system has been suggested by Arenas et al. [37]. They study the time development of the synchronization process of phase-coupled so-called Kuramoto oscillators. Oscillators are placed on the nodes on the networks and initialized with random phases. The couplings are determined via the links of the network. Densely interconnected groups of oscillators tend to synchronize first and therefore the community structure and its hierarchy can be inferred from studying the matrix of phase correlations as the system progresses from a completely uncorrelated to a completely correlated state when all oscillators are in phase.

The idea to study a dynamical system on a data set for clustering purposes was first introduced by Blatt et al. [38–40]. They studied the spin–spin correlation of a ferromagnetic Potts model during a simulated cooling process from the paramagnetic state to the complete ferromagnetic state. Clusters are interpreted as groups of spins which are highly correlated.

Both of the algorithms based on dynamical systems may be run on large systems as the interactions are defined only along the sparse connections of the network. However, they require a full $N \times N$ correlation matrix to be kept in memory which makes them impractical for very large systems.

In total, all hierarchical algorithms suffer from the fact that a community needs to be understood as something output by the algorithm and hence the definition of what a community is depends on the particular choice of rule to remove an edge, join nodes to communities or on the dynamical system studied. One may use an external definition of community in order to decide where to best cut the dendrogram. Then, however, it is not clear whether the algorithm chosen really does optimize this quality measure. Further, all hierarchical algorithms imply the existence of a community structure at all levels of detail from single nodes to the whole network. There is no true interpretation of overlap other than what results from ad hoc introduced sampling procedures.

2.4.3 Semi-hierarchical

The hierarchical methods cited so far assume a nested hierarchy of communities. One of the few methods which allow for overlapping communities is the clique percolation method of Palla et al. [8, 22] which was introduced already. Even though the method allows a node to be part of more than one community, communities resulting from $k + 1$-clique percolation processes are always contained within k-clique communities. It is never possible that the nodes contained in the overlap of two communities form their own community. Another problem of this method is its dependence on the existence of triangles in the network. Nodes which are not connected via triangles to communities can never be part of such communities and only nodes with at least $k - 1$ links can be part of a k-clique at all. Also, this method may be easily mislead by the addition or removal of single links in the network, as a single link may be responsible for the joining of two communities into one. Clearly, this situation is unsatisfactory in case of noisy data.

2.4.4 Non-hierarchical

The non-hierarchical methods approach the problem from a different perspective. In principle, they intend to calculate a full distance matrix for the nodes of the network. This can then be treated by conventional techniques.

One of the earliest approaches to community detection is due to Eriksen et al. [41, 42]. They study a diffusion process on a network and analyze the decay of the modes of the following diffusive system with discrete time:

$$\rho_i(t+1) - \rho_i(t) = \sum_j (T_{ij} - \delta_{ij})\rho_j(t). \tag{2.6}$$

Here T_{ij} represents the adjacency matrix of the network such that $T_{ij} = 1/k_j$ for $A_{ij} = 1$ and zero otherwise. Hence T_{ij} represents the probability of a random walker to go from j to i. The decay of a random initial configuration $\rho(t = 0)$ toward the steady state is characterized by the eigenmodes of the transition matrix T_{ij}. The eigenvectors corresponding to the largest eigenvalues can then be used to define a distance between nodes which helps in identifying communities. To do this, the eigenvectors belonging to the largest non-trivial positive eigenvalues are plotted against each other. This diffusion approach is very similar in spirit to other algorithms based on the idea of using flow simulations for community detection as suggested by van Dongen [43] under the name of "Markov clustering" (MCL).

The method presented by Zhou [44–46] first converts the sparse adjacency matrix of the graph into a full distance matrix by calculating the average time a Brownian particle needs to move from node i to j. Then this distance matrix is clustered using ordinary hierarchical clustering algorithms. This approach is based on the observation that a random walker has shorter traveling time between two nodes if many (short) alternative paths exist.

Another spectral approach has been taken by Muños and Donetti [47]. They work with the Laplacian matrix of the network. The Laplacian is defined as

$$L_{ij} = k_i\delta_{ij} - A_{ij}. \tag{2.7}$$

Otherwise, the method proposed is similar to Ref. [41]. Plotting the non-trivial eigenvectors against each other gives a low-dimensional representation of a distance measure of the network on top of which a conventional clustering procedure then needs to operate.

Though these methods are able to recover known community structures with good accuracy, they suffer from being less intuitive. Communities found can only be interpreted with respect to the particular system under study, be it a diffusive system or the eigen vectors of the Laplacian matrix. Problematic is also that there is no local variant of these methods, i.e., there is no way to find the community around a given node using spectral methods.

2.4.5 Optimization Based

A different approach which is reminiscent of the parametric clustering procedures known in computer science is the idea of searching for partitions with maximum modularity Q using combinatorial optimization techniques [48]. This approach has been adopted by Guimera et al. in Refs. [2, 49] or Massen et al. [50] using simulated annealing [51] or Duch and Arenas using extremal optimization [52].

Though this approach will be the preferred one for the remainder of this book, a number of issues remain. For the hierarchical algorithms, a community was to be understood as whatever the algorithm outputs. Now, it is not the algorithm that defines what a community is, but the quality function, i.e., the modularity Q in this case. Also, the modularity Q as defined by Newman [23] is parameter free and an understanding for hierarchical and overlapping structures needs to be developed.

2.5 Conclusion

Block structure in networks is a very common and well-studied phenomenon. The concepts of structural and regular equivalence as well as the types of blocks defined for generalized block modeling are well defined but appear too rigid to be of practical use for large and noisy data sets. Diagonal block models or modular structures have received particular attention in the literature and have developed into an almost independent concept of cohesive subgroups or communities. The comparison of many different community definitions from various fields has shown that the concept of module or community in a network is only vaguely defined. The diversity of algorithms published is only a consequence of this vague definition. None of the algorithms could be called "ideal" in the sense that it combines the features of computational efficiency, accuracy, flexibility and adaptability with regard to the network and easy interpretation of the results. More importantly, none of the above-cited publications allows an estimation to which degree the community structure found is a reality of the network or a product of the clustering process itself. The following chapters are addressing these issues and present a framework in which community detection is viewed again as a special case of a general procedure for detecting block structure in networks.

References

1. M. E. J. Newman. The structure and function of complex networks. *SIAM Review*, 45(2):167–256, 2003.
2. R. Guimera and L. A. N. Amaral. Functional cartography of complex metabolic networks. *Nature*, 433:895–900, 2005.
3. H. Jeong, B. Tombor, R. Albert, Z. N. Oltvai, and A. -L. Barabàsiasi. The large-scale organization of metabolic networks. *Nature*, 407:651–654, 2000.
4. A. -L. Barabási and Z. N. Oltvai. Network biology: Understanding the cells's functional organization. *Nature Reviews Genetics*, 5:101–113, 2004.
5. H. Jeong, S. Mason, A. -L. Barabàsi, and Z. N. Oltvai. Lethality and centrality in protein networks. *Nature*, 41:41–42, 2001.
6. R. Guimera, M. Sales-Pardo, and L. A. N. Amaral. Classes of complex networks. *Nature Physics*, 3:63–69, 2007.
7. D. M. Wilkinson and B. A. Huberman. A method for finding communities of related genes. *Proceedings of the National Academy of Sciences of the United States of America*, 101:5241–5248, 2004.

8. G. Palla, I. Derenyi, I. Farkas, and T. Vicsek. Uncovering the overlapping community structure of complex networks in nature and society. *Nature*, 435:814, 2005.
9. J. Reichardt and S. Bornholdt. Clustering of sparse data via network communities – a prototype study of a large online market. *Journal of Statistical Mechanics*, P06016, 2007.
10. S. P. Borgatti and M. G. Everett. Notions of position in social network analysis. *Sociological Methodology*, 22:1–35, 1992.
11. S. Wasserman and K. Faust. *Social Network Analysis*. Cambridge University Press, New York, 1994.
12. F. Lorrain and H. C. White. Structural equivalence of individuals in social networks. *The Journal of mathematical sociology*, 1:49–80, 1971.
13. D. R. White and K. P. Reitz. Graph and semigroup homomorphisms. *Social Networks*, 5:193–234, 1983.
14. M. G. Everett and S. P. Borgatti. Regular equivalence: general theory. *The Journal of Mathematical Sociology*, 19:29–52, 1994.
15. P. Doreian, V. Batagelj, and A. Ferligoj. *Generalized Blockmodeling*. Cambridge University Press, New York, 2005.
16. M. Mcperson, L. Smith-Lovin, and J. M. Cook. Birds of a feather: Homophily in social networks. *Annual Review of Sociology*, 27:415:44, 2001.
17. M. Girvan and M. E. J. Newman. Community structure in social and biological networks. *Proceedings of the National Academy of Sciences of the United States of America*, 99(12):7821–7826, 2002.
18. W. Zachary. An information flow model for conflict and fission in small groups. *Journal of Anthropological Research*, 33:452–473, 1977.
19. M. E. J. Newman. Detecting community structure in networks. *European Physical Journal B*, 38:321, 2004.
20. L. Danon, J. Dutch, A. Arenas, and A. Diaz-Guilera. Comparing community structure indentification. *Journal of Statistical Mechanics*, P09008, 2005.
21. F. Radicchi, C. Castellano, F. Cecconi, V. Loreto, and D. Parisi. Defining and identifying communities in networks. *Proceedings of the National Academy of Sciences of the United States of America*, 101:2658, 2004.
22. I. Derényi, G. Palla, and T. Vicsek. Clique percolation in random networks. *Physical Review Letters*, 94:160202, 2005.
23. M. E. J. Newman and M. Girvan. Finding and evaluating community structure in networks. *Physical Review E*, 69:026113, 2004.
24. L. Kaufman and P. J. Rousseeuw. *Finding Groups in Data:An Introduction to Cluster Analysis*. Wiley-Interscience, New York, 1990.
25. B.S. Everitt, S. Landau, and M. Leese. *Cluster Analysis*. Arnold, London, 4 edition, 2001.
26. A. K. Jain, M. N. Murty, and P. J. Flynn. Data clustering: A review. *ACM Computing Surveys*, 31(3):264–323, 1999.
27. B. Kernighan and S. Lin. An effective heuristic procedure for partitioning graphs. *The Bell System Technical Journal*, 29:291–307, 1970.
28. C. Fiduccia and R. Mattheyses. A linear time heuristic for improving network partitions. In *Proceedings of the 19th Design Automation Confrence*, pp. 175–181, 1982.
29. C. -K. Cheng and Y. A. Wei. An improved two-way partitioning algorithm with stable performance. *IEEE Transactions on Computer-Aided Design, Integrated Circuits Systems*, 10:1502–1511, 1991.

30. L. Hagen and A. B. Kahng. New spectral methods for ratio cut partitioning and clustering. *IEEE Transactions on Computer-Aided Design*, 11:1074–1085, 1992.

31. J. Shi and J. Malik. Normalized cuts and image segmentation. *IEEE Transactions Pattern Analysis and Machine Intelligence*, 22(8):888–905, 2000.

32. Ch. H. Q. Ding, X. He, H. Zha, M. Gu, and H. D. Simon. A min-max cut algorithm for graph partioning and data clustering. In *Proceedings of ICDM 2001*, pp. 107–114, 2001.

33. J. R. Tyler, D. M. Wilkinson, and B. A. Huberman. Email as spectroscopy: automated discovery of community structure within organisations. In *International Conference on Communities and Technologies*, Amsterdam, The Netherlands, 2003.

34. M. E. J. Newman. A measure of betweenness centrality based on random walks. Social Networks, 27(1), 39–54, 2005.

35. F. Wu and B. A. Huberman. Finding communities in linear time: a physics approach. *European Physical Journal B*, 38:331, 2004.

36. M. E. J. Newman. Fast algorithm for detecting community structure in networks. *Physical Review E*, 69:066133, 2004.

37. A. Arenas, A. D'iaz-Guilera, and C. J. Pérez-Vicente. Synchronization reveals topological scales in complex networks. *Physical Review Letters*, 96:114102, 2006.

38. M. Blatt, S. Wiseman, and E. Domany. Super-paramagnetic clustering of data. *Physical Review Letters*, 76:3251–3254, 1996.

39. S. Wiseman, M. Blatt, and E. Domany. Super-paramagnetic clustering of data. Physical Review E, 57, 3767, 1998.

40. E. Domany. Cluster analysis of gene expression data. *Journal of Statistical Mechanics*, 110(3–6):1117–1139, 2003.

41. K. A. Eriksen, I. Simonsen, S. Maslov, and K. Sneppen. Modularity and extreme edges of the internet. *Physical Review Letters*, 90(14), 148701, 2003.

42. I. Somonsen, K. A. Eriksen, S. Maslov, and K. Sneppen. Diffusion on complex networks: a way to probe their large scale topological structures. *Physica A*, 336:163, 2004.

43. S. van Dongen. *Graph Clustering by Flow Simulation*. PhD thesis, University of Utrecht, The Netherlands, 2000.

44. H. Zhou. Distance, dissimilarity index, and network community structure. *Physical Review E*, 67: 067907, 2003

45. H. Zhou and R. Lipowsky. *Network Brownian Motion: A New Method to Measure Vertex-Vertex Proximity an to Identify Communities and Subcommunities*, pp. 1062–1069. Springer-Verlag, Berlin Heidelberg, 2004.

46. H. Zhou. Network landscape from a brownian particle's perspective. *Physical Review E*, 67:041908, 2003.

47. L. Donetti and M. A. Munoz. Detecting network communities: a new and systematic approach. *Journal of Statistical Mechanics: Theory and Experiment*, P10012, 2004.

48. C. H. Papadimitriou. *Combinatorial Optimization: Algorithms and Complexity*. Dover Publications, New York, 1998.

49. R. Guimera, M. Sales-Pardo, and L. N. Amaral. Modularity from fluctuations in random graphs and complex networks. *Physical Review E*, 70:025101(R), 2004.

50. C. P. Massen and J. P. K. Doye. Identifying communities within energy landscapes. *Physical Review E*, 71:046101, 2005.
51. S. Kirkpatrick, C.D. Gelatt Jr., and M.P. Vecchi. Optimization by simulated annealing. *Science*, 220:671–680, 1983.
52. J. Duch and A. Arenas. Community detection in complex networks using extremal optimization. *Physical Review E*, 72:027104, 2005.

3

A First Principles Approach to Block Structure Detection

3.1 Mapping the Problem

Common to all of the before-mentioned approaches is their attempt to discover patterns in the link structure of networks. Patterns were either block structures in the adjacency matrix or – more specifically – cohesive subgroups. We will try to define a quality function for block structure in networks and optimize the ordering of rows and columns of the matrix as to maximize the quality of the blocking. The search for cohesive subgroups will prove to be a special case of this treatment. It makes sense to require that our quality function will be independent of the order of rows and columns within one block. It will depend only on the assignment of nodes, i.e., rows and columns, into blocks. Finding a good assignment into blocks is hence a combinatorial optimization problem. In many cases, it is possible to map such a combinatorial optimization problem onto minimizing the energy of a spin system [1]. This approach had been suggested for the first time by Fu and Anderson in 1986 [2] in the context of bi-partitioning of graphs and it has been applied successfully to other problems such as vertex cover [3], k-sat [4], the traveling salesmen [5] and many others as well.

Before introducing such a quality function, it is instructive to leave the field of networks for a moment and take a detour into the dimensionality reduction of multivariate data.

3.1.1 Dimensionality Reduction with Minimal Squared Error

Suppose we are given a set of real valued measurements of some objects. As an example, for all boats in a marina, we measure length over all, width, height of the mast, the area of the sail, power of the engine, length of the waterline, and so forth. Let N be the number of measurements, i.e., the number of boats in the marina, and let the measurements be vectors of dimension d, i.e., the number of things we have measured. We compile our measurements into a

Reichardt, J.: *A First Principles Approach to Block Structure Detection*. Lect. Notes Phys. **766**, 31–43 (2009)
DOI 10.1007/978-3-540-87833-9_3   © Springer-Verlag Berlin Heidelberg 2009

data matrix $\mathbf{A} \in \mathbb{R}^{N \times d}$, i.e., we write the individual measurement vectors as the rows of matrix $\mathbf{A}$. Let us further assume that we have already subtracted the mean across all measurements from each individual sample such that the columns of $\mathbf{A}$ sum to zero, i.e., we have centered our data.

Now we see that $\mathbf{A}^T\mathbf{A}$ is a $d \times d$ matrix describing the covariance of the individual dimensions in which we measured our data.

We now ask if we can drop some of the d dimensions and still describe our data well. Naturally, we want to drop those dimensions in which our data do not vary much or we would like to replace two dimensions which are correlated by a single dimension. We can discard the unnecessary dimensions by projecting our data from the d-dimensional original space in a lower dimensional space of dimension $q < d$. Such a projection can be achieved by a matrix $\mathbf{V} \in \mathbb{R}^{d \times q}$. Taking measurement $\mathbf{a_i} \in \mathbb{R}^d$ from row i of $\mathbf{A}$, we find the coordinates in the new space to be $\mathbf{b_i} = \mathbf{a_i V}$ with $\mathbf{b_i} \in \mathbb{R}^q$. We can also use the transpose of $\mathbf{V}$ to project back into the original space of dimension d via $\mathbf{a_i'} = \mathbf{b_i V}^T$. Since in the two projections we have visited a lower dimensional space, we find that generally the reconstructed data point does not coincide with the original datum $\mathbf{a_i VV}^T = \mathbf{a_i'} \neq \mathbf{a_i}$.

However, if we would have first started in the q-dimensional space with $\mathbf{b_i}$ and projected it into the d-dimensional space via $\mathbf{V}^T$ and then back again via $\mathbf{V}$ we require that our projection does not lose any information and hence $\mathbf{b_i V}^T\mathbf{V} = \mathbf{b_i}$. This means that we require $\mathbf{V}^T\mathbf{V} = \mathbb{1}$ or in other words we require that our projection matrix $\mathbf{V}$ be unitary.

The natural question is now how to find a unitary matrix such that it minimizes some kind of reconstruction error. Using the mean square error, we could write

$$E \propto \sum_i^N \sum_j^d (\mathbf{A} - \mathbf{A}')_{ij}^2 = \sum_i^N \sum_j^d (\mathbf{A} - \mathbf{AVV^T})_{ij}^2 \tag{3.1}$$

$$= \mathrm{Tr}(\mathbf{A} - \mathbf{AVV}^T)^T(\mathbf{A} - \mathbf{AVV}^T). \tag{3.2}$$

The new coordinates that we project our data onto are called "principal components" of the data set and the technique of finding them is known as "principal component analysis" or PCA for short. Already at this point, we can mention that the q columns of $\mathbf{V}$ must be made of the eigenvectors belonging to the largest q eigenvalues of $\mathbf{A}^T\mathbf{A}$. To show this, we discuss a slightly different problem, solve it and then show that it is equivalent to the above.

Consider the singular value decomposition (SVD) of a matrix of $\mathbf{A} \in \mathbb{R}^{N \times d}$ into a unitary matrix $\mathbf{U} \in \mathbb{R}^{N \times N}$, a diagonal matrix $\mathbf{S} \in \mathbb{R}^{N \times d}$ (in case $N \neq d$ there are maximally $\min(N, d)$ non-zero entries, the number of non-zero entries in $\mathbf{S}$ is the rank of $\mathbf{A}$) and another unitary matrix $\mathbf{V} \in \mathbb{R}^{d \times d}$ such that $\mathbf{A} = \mathbf{USV}^T$ and $\mathbf{S} = \mathbf{U}^T\mathbf{AV}$. The entries on the diagonal of $\mathbf{S}$ are called singular values. We will assume that they are ordered decreasing in absolute value. It is straightforward to see some of the properties of this SVD: $\mathbf{U}^T\mathbf{A} = \mathbf{SV}^T$ and $\mathbf{AV} = \mathbf{US}$ follow from the $\mathbf{U}$ and $\mathbf{V}$ being unitary.

In the same fashion, we have $\mathbf{A}\mathbf{A}^T = \mathbf{U}\mathbf{S}\mathbf{S}^T\mathbf{U}^T$ and $\mathbf{A}^T\mathbf{A} = \mathbf{V}\mathbf{S}^T\mathbf{S}\mathbf{V}^T$ which means that $\mathbf{U}$ diagonalizes $\mathbf{A}\mathbf{A}^T$ and $\mathbf{V}$ diagonalizes $\mathbf{A}^T\mathbf{A}$ with the *same* eigenvalues.

Now consider the matrix $\mathbf{A}' = \mathbf{U}\mathbf{S}'\mathbf{V}^T$ where $\mathbf{S}'$ differs from $\mathbf{S}$ in that we have kept only the q largest (in absolute value) entries in $\mathbf{S}$. That is, we construct a rank q approximation to $\mathbf{A}$. How does $\mathbf{A}'$ differ from $\mathbf{A}$? The quadratic error now reads

$$E \propto \sum_i^N \sum_j^d (\mathbf{A} - \mathbf{A}')_{ij}^2 = \mathrm{Tr}(\mathbf{A} - \mathbf{A}')^T(\mathbf{A} - \mathbf{A}') \tag{3.3}$$

$$= \mathrm{Tr}(\mathbf{U}^T\mathbf{A}\mathbf{V} - \mathbf{U}^T\mathbf{A}'\mathbf{V})^T(\mathbf{U}^T\mathbf{A}\mathbf{V} - \mathbf{U}^T\mathbf{A}'\mathbf{V})$$

$$= \mathrm{Tr}(\mathbf{S} - \mathbf{S}')^T(\mathbf{S} - \mathbf{S}')$$

$$= \sum_{r=q+1}^{\min(N,d)} \mathbf{S}_{rr}^2. \tag{3.4}$$

Trivially, choosing $q = \min(N, d)$ would make this error zero. However, it can be shown that this choice of constructing $\mathbf{A}'$ is also the best possible choice of approximating $\mathbf{A}$ by a matrix of rank q under the squared error function.

Now going back to PCA, we see that the term $\mathbf{A}\mathbf{V}\mathbf{V}^T$ in (3.2) corresponds just to $\mathbf{A}\mathbf{V}\mathbf{V}^T = \mathbf{U}\mathbf{S}'\mathbf{V}^T$ from the SVD which means that the minimal reconstruction error is obtained via the $d \times q$ matrix $\mathbf{V}$, the columns of which are the right singular vectors of $\mathbf{A}$ corresponding to the q largest singular values or the eigenvectors of $\mathbf{A}^T\mathbf{A}$ corresponding to the q largest eigenvalues.

What we learn from these considerations is that very simple and well understood algebraic techniques exist to minimize the squared error when reducing the dimensionality of a multivariate data set. We will argue, however, that this is not necessarily a good error function for networks and will hence try to give a better one. The price we will have to pay for this will be a higher computational cost for minimizing this error.

3.1.2 Squared Error for Multivariate Data and Networks

Let us consider in the following the reconstruction of the adjacency matrix of a network $\mathbf{A} \in \{0, 1\}^{N \times N}$ of rank r by another adjacency matrix $\mathbf{B} \in \{0, 1\}^{N \times N}$ possibly of lower rank $q < r$ as before. For the squared error we have

$$E = \sum_{ij}(\mathbf{A} - \mathbf{B})_{ij}^2. \tag{3.5}$$

Then, there are only four different cases we need to consider in Table 3.1. The squared error gives equal value to the mismatch on the edges and missing edges in $\mathbf{A}$. We could say it weighs every error by its own magnitude. While

Table 3.1. The error matrix of the quadratic error (3.5). Each error is weighted by its own magnitude. Making a mistake in matching an edge in **A** is as bad as mismatching a missing edge in **A**.

A_{ij} \ B_{ij}	1	0
1	0	1
0	1	0

this is a perfectly legitimate approach for multivariate data, it is, however, highly problematic for networks. The first reason is that many networks are sparse. The fraction of non-zero entries in **A** is generally very, very small compared to the fraction of zero entries. A low rank approximation under the squared error will retain this sparsity to the point that **B** may be completely zero. Furthermore, we have seen that real networks tend to have a very heterogeneous degree distribution, i.e., the distribution of zeros and ones per row and column in **A** is also very heterogeneous. Why give every entry the same weight in the error function? Most importantly, for multivariate data, all entries of $\mathbf{A}_{ij}$ are equally important measurements in principle. For networks this is not the case: the edges are in principle more important than the missing edges. There are fewer of them and they should hence be given more importance than missing edges. Taken all of these arguments together, we see that our first goal will have to be the derivation of an error function specifically tailored for networks that does not suffer from these deficiencies.

3.2 A New Error Function

We already said that we would like to use a statistical mechanics approach. The problem of finding a block structure which reflects the network as good as possible is then mapped onto finding the solution of a combinatorial optimization problem. Trying to approximate the adjacency matrix **A** of rank r by a matrix **B** of rank $q < r$ means approximating **A** with a block model of only full and zero blocks. Formally, we can write this as $\mathbf{B}_{ij} = B(\sigma_i, \sigma_j)$ where $B(r, s)$ is a $\{0, 1\}^{q \times q}$ matrix and $\sigma_i \in \{1, ..., q\}$ is the assignment of node i from **A** into one of the q blocks. We can view $B(r, s)$ as the adjacency matrix of the blocks in the network or as the image graph discussed in the previous chapter and its nodes represent the different equivalence classes into which the vertices of **A** may be grouped. From Table 3.1, we see that our error function can have only four different contributions. They should

1. reward the matching of edges in **A** to edges in **B**,
2. penalize the matching of missing edges (non-links) in **A** to edges in **B**,
3. penalize the matching of edges in **A** to missing edges in **B** and
4. reward the matching of missing edges in **A** to edges in **B**

These four principles can be expressed via the following function:

$$\mathcal{Q}(\{\sigma\}, \mathbf{B}) = \sum_{ij} a_{ij} \underbrace{A_{ij} B(\sigma_i, \sigma_j)}_{\text{links to links}} - \sum_{ij} b_{ij} \underbrace{(1 - A_{ij}) B(\sigma_i, \sigma_j)}_{\text{non-links to links}}$$

$$- \sum_{ij} c_{ij} \underbrace{A_{ij}(1 - B(\sigma_i, \sigma_j))}_{\text{links to non-links}} + \sum_{ij} d_{ij} \underbrace{(1 - A_{ij})(1 - B(\sigma_i, \sigma_j))}_{\text{non-links to non-links}},$$

(3.6)

in which A_{ij} denotes the adjacency matrix of the graph with $A_{ij} = 1$, if an edge is present and zero otherwise, $\sigma_i \in \{1, 2, ..., q\}$ denotes the role or group index of node i in the graph and $a_{ij}, b_{ij}, c_{ij}, d_{ij}$ denote the weights of the individual contributions, respectively. The number q determines the maximum number of groups allowed and can, in principle, be as large as N, the number of nodes in the network. Note that in an optimal assignment of nodes into groups it is not necessary to use all group indices as some indices may remain unpopulated in the optimal assignment.

We will not restrict our analysis to a particular type of network. If the network is directed, the matrix $\mathbf{A}$ is asymmetric. If the network is weighted, we assume $\mathbf{A}$ to represent the $\{0, 1\}$ adjacency structure and $\mathbf{w} \in \mathbb{R}_+^{N \times N}$ to hold the weights of the links in $\mathbf{A}$. Naturally, we have $w_{ij} = A_{ij}$ in case of unweighted networks. The extension to bipartite networks, i.e., adjacency matrices $\mathbf{A}$ which are not square anymore, is straightforward as well.

In principle, (3.6) is formally equivalent to the Hamiltonian of a q-state Potts model [6]. However, the spin interaction is governed by $B(\sigma_i, \sigma_j)$ which is more general than the standard Potts model $B(\sigma_i, \sigma_j) = \delta(\sigma_i, \sigma_j)$. The couplings between the spins are derived from the (weighted) adjacency matrix of the graph. The spin state of a node serves as a block index, such that nodes in the same spin state belong to the same block. The ground state, or the spin configuration with minimal energy, will then be equivalent to an optimal assignment of nodes into blocks according to the error function.

It is natural to weigh the links and non-links in $\mathbf{A}$ equally, regardless of whether they are mapped onto edges or missing edges in $\mathbf{B}$, i.e., $a_{ij} = c_{ij}$ and $b_{ij} = d_{ij}$. It remains to find a sensible choice of weights a_{ij} and b_{ij}, preferably such that the contribution of links and non-links can be adjusted through a parameter. The a_{ij} measure the contribution of the matching of edges while the b_{ij} measure the contribution of the matching of missing edges. From our discussion of the squared error, we have seen that this should somehow compensate the sparsity of the networks. A convenient way to achieve this is setting $a_{ij} = w_{ij} - b_{ij}$. Then, a natural condition is that the *total* amount of "quality" that can possibly be contributed by links and non-links, respectively, should be equal. In other words $\sum_{ij} A_{ij} a_{ij} = \sum_{ij}(1 - A_{ij}) b_{ij}$. This also means that $\sum_{ij} w_{ij} A_{ij} = \sum_{ij} b_{ij}$. In case we would like to tune the influence of edges and missing edges by a parameter γ, it is convenient to introduce it as $b_{ij} = \gamma p_{ij}$ with the restriction that $\sum_{ij} w_{ij} A_{ij} = \sum_{ij} p_{ij}$. Here, we have introduced p_{ij} merely as a penalty we give to matching missing edges in $\mathbf{A}$ to edges in $\mathbf{B}$. However, from $\sum_{ij} w_{ij} A_{ij} = \sum_{ij} p_{ij}$ we may also interpret p_{ij} as a measure for the probability that nodes i and j are connected or – in general – for the expected weight between them. We will discuss this point further later on. Finally, for parameter values of $\gamma = 1$ we give equal total weights to edges and missing edges, whereas values of γ smaller or greater than one give more total weight to edges or missing edges, respectively. Then we can write (3.6) as

$$\mathcal{Q}(\{\sigma\}, \mathbf{B}) = \sum_{ij} (w_{ij} A_{ij} - \gamma p_{ij}) B(\sigma_i, \sigma_j)$$

$$- \sum_{ij} (w_{ij} A_{ij} - \gamma p_{ij})(1 - B(\sigma_i, \sigma_j)). \tag{3.7}$$

In (3.7) we note that the term $\sum_{ij}(w_{ij} A_{ij} - \gamma p_{ij})$ does not depend on the block model $B(r, s)$ or on the assignment of nodes into blocks $\{\sigma\}$. Hence, the matrix $B(r, s)$ and the assignment of $\{\sigma\}$ which maximize the first term of (3.7) will also minimize the second. It is thus enough to optimize only one of the two terms.

Let us derive our quality function in a different way. Similar to (3.5) we could write as an error function

$$E = \sum_{ij}^{N} (\mathbf{A} - \mathbf{B})_{ij} (\mathbf{w} - \gamma \mathbf{p})_{ij} \tag{3.8}$$

$$= \sum_{ij}^{N} (\mathbf{w} - \gamma \mathbf{p})_{ij} A_{ij} - \sum_{ij}^{N} (\mathbf{w} - \gamma \mathbf{p})_{ij} B(\sigma_i, \sigma_j). \tag{3.9}$$

$$\underbrace{\qquad\qquad\qquad}_{\text{Optimal Fit}} \underbrace{\qquad\qquad\qquad\qquad}_{\text{Approximate Fit}}$$

$$\tag{3.10}$$

Note how the different types of errors are weighted differently. Compare Table 3.2 with Table 3.1 to emphasize this difference again. We see immediately that the second part of the error function (3.10) corresponds to the first part of our quality function (3.7). We can interpret the error as a difference of an optimal fit achieved when $A_{ij} = B(\sigma_i, \sigma_j)$ and the approximate

Table 3.2. The error matrix of the linear error (3.10). Each type of error is weighted by its own weight. Making a mistake in matching an edge in $\mathbf{A}$ is *worse* than mismatching a missing edge in $\mathbf{A}$.

A_{ij} \ B_{ij}	1	0
1	0	$w_{ij} - \gamma p_{ij}$
0	γp_{ij}	0

fit that we achieve for a given $B(r,s)$ and assignment of nodes into groups σ_i. It is worth noting that both $B(r,s) = 1$ and $B(r,s) = 0$ for all r,s lead to the same error value for $\gamma = 1$. Further, the error function is maximal, if $B_{\sigma_i,\sigma_j} = 1 - A_{ij}$, i.e., exactly complementary to A_{ij}.

3.2.1 Fitting Networks to Image Graphs

The above-defined quality and error functions in principle consist of two parts. On one hand, there is the image graph $\mathbf{B}$ and on the other hand, there is the mapping of nodes of the network to nodes in the image graph, i.e., the assignment of nodes into blocks, which both determine the fit. Given a network $\mathbf{A}$ and an image graph $\mathbf{B}$, we could now proceed to optimize the assignment of nodes into groups $\{\sigma\}$ as to optimize (3.6) or any of the derived forms. This would correspond to "fitting" the network to the given image graph. This allows us to compare how well a particular network may be represented by a given image graph. We will see later that the search for cohesive subgroups is exactly of this type of analysis: If our image graph is made of isolated vertices which only connect to themselves, then we are searching for an assignment of nodes into groups such that nodes in the same group are as densely connected as possible and nodes in different groups as sparsely as possible. However, ultimately, we are interested also in the image graph which best fits to the network among all possible image graphs $\mathbf{B}$. In principle, we could try out every possible image graph, optimize the assignment of nodes into blocks $\{\sigma\}$ and compare these fit scores. This quickly becomes impractical for even moderately large image graphs. In order to solve this problem, it is useful to consider the properties of the optimally fitting image graph $\mathbf{B}$ if we are given the networks plus the assignment of nodes into groups $\{\sigma\}$.

3.2.2 The Optimal Image Graph

We have already seen that the two terms of (3.7) are extremized by the same $B(\sigma_i, \sigma_j)$. It is instructive to introduce the abbreviations

$$m_{rs} = \sum_{ij} w_{ij} A_{ij} \delta(\sigma_i, r)\delta(\sigma_j, s) \text{ and} \tag{3.11}$$

$$[m_{rs}]_{p_{ij}} = \sum_{ij} p_{ij} \delta(\sigma_i, r)\delta(\sigma_j, s), \tag{3.12}$$

and write two equivalent formulations for our quality function:

$$Q^1(\{\sigma\}, \mathbf{B}) = \sum_{r,s} \left(m_{rs} - \gamma[m_{rs}]_{p_{ij}} \right) B(r,s) \text{ and} \tag{3.13}$$

$$Q^0(\{\sigma\}, \mathbf{B}) = -\sum_{r,s} \left(m_{rs} - \gamma[m_{rs}]_{p_{ij}} \right) (1 - B(r,s)). \tag{3.14}$$

Now the sums run over the group indices instead of nodes and m_{rs} denotes the number of edges between nodes in group r and s and $[m_{rs}]_{p_{ij}}$ is the sum of penalties between nodes in group r and s. Interpreting p_{ij} indeed as a probability or expected weight, the symbol $[\cdot]_{p_{ij}}$ denotes an expectation value under the assumption of a link(weight) distribution p_{ij}, given the current assignment of nodes into groups. That is, $[m_{rs}]_{p_{ij}}$ is the expected number (weight) of edges between groups r and s. The equivalence of maximizing (3.13) and minimizing (3.14) shows that our quality function is insensitive to whether we optimize the matching of edges or missing edges between the network and the image graph.

Let us now consider the properties of an image graph with q roles and a corresponding assignment of roles to nodes which would achieve the highest Q across all image graphs with the same number of roles. From (3.13) and (3.14) we find immediately that for a given assignment of nodes into blocks $\{\sigma\}$ we achieve that Q is maximal only when $B_{rs} = 1$ for every $(m_{rs} - [m_{rs}]) > 0$ and $B_{rs} = 0$ for every $(m_{rs} - [m_{rs}]) < 0$. This means that for the best fitting image graph, we have more links than expected between nodes in roles connected in the image graph. Further, we have less links than expected between nodes in roles disconnected in the image graph.

This suggests a simple way to eliminate the need for a given image graph by considering the following quality function:

$$Q(\{\sigma\}) = \frac{1}{2} \sum_{r,s} \|m_{rs} - \gamma[m_{rs}]\|. \tag{3.15}$$

The factor $1/2$ enters to make the scores of Q, Q^0 and Q^1 comparable. From the assignment of roles that maximizes (3.15), we can read off the image graph simply by setting

$$B_{rs} = 1, \text{ if } (m_{rs} - \gamma[m_{rs}]) > 0 \text{ and} \tag{3.16}$$
$$B_{rs} = 0, \text{ if } (m_{rs} - \gamma[m_{rs}]) \le 0. \tag{3.17}$$

3.2.3 Maximum Value of the Fit Score

The function (3.15) is monotonously increasing with the number of possible roles q until it reaches its maximum value $Q_{\max}$

$$Q_{\max} = \sum_{ij} (w_{ij} A_{ij} - \gamma p_{ij}) A_{ij}. \tag{3.18}$$

This value can be achieved when q equals the number of structural equivalence classes in the network, i.e., the number of rows/columns which are genuine in $\mathbf{A}$. The optimal assignment of roles $\{\sigma\}$ is then simply an assignment into the structural equivalence classes. For fewer roles, this allows us to compare $Q/Q_{\max}$ for the actual data and a randomized version and to use this comparison as a basis for the selection of the optimal number of roles in the image graph in order to avoid overfitting the data.

A comparison of the image graphs and role assignments found independently for different numbers of roles may also allow for the detection of possible hierarchical or overlapping organization of the role structure in the network.

3.2.4 Choice of a Penalty Function and Null Model

We have introduced p_{ij} as a penalty on the matching of missing links in $\mathbf{A}$ to links in $\mathbf{B}$. As such, it can in principle take any form or value that may seem suitable. However, we have already hinted at the fact that p_{ij} can also be interpreted as a probability. As such, it provides a random null model for the network under study. The quality functions (3.13), (3.13) and (3.15) then all compare distribution of links as found in the network for a given assignment of nodes into blocks to the expected link (weight) distribution if links (weight) were distributed independently of the assignment of nodes into blocks according to p_{ij}. Maximizing the quality functions (3.13), (3.13) and (3.15) hence means to find an assignment of nodes into blocks such that the number (weight) of edges in blocks deviates as strongly as possible from the expectation value due to the random null model.

Two exemplary choices of link distributions or random null models shall be illustrated. Both fulfill the constraint that $\sum_{ij} w_{ij} A_{ij} = \sum_{ij} p_{ij}$. The simplest choice is to assume every link equally probable with probability $p_{ij} = p$ independent from i to j. Writing

$$p_{ij} = p = \frac{\sum_{kl} w_{kl} A_{kl}}{N^2} \tag{3.19}$$

leads naturally to

$$[m_{rs}]_p = p n_r n_s, \tag{3.20}$$

with n_r and n_s denoting the number of nodes in group r and s, respectively.

A second choice for p_{ij} may take into account that the network does exhibit a particular degree distribution. Since links are in principle more likely between nodes of high degree, matching links between high-degree nodes should get a lower reward and mismatching them a higher penalty. One may write

$$p_{ij} = \frac{(\sum_k w_{ik} A_{ik})(\sum_l w_{lj} A_{lj})}{\sum_{kl} w_{kl} A_{kl}} = \frac{k_i^{out} k_j^{in}}{M}, \tag{3.21}$$

which takes this fact and the degree distribution into account. In this form, the penalty p_{ij} is proportional to the product of the row and column sums of the weight matrix $\mathbf{w}$. The number (weight) of outgoing links of node i is given by k_i^{out} and the number (weight) of incoming links of node j is given by k_j^{in}. With these expressions one can write

$$[m_{rs}]_{p_{ij}} = \frac{1}{M} K_r^{out} K_s^{in}. \tag{3.22}$$

Here, K_r^{out} is the sum of weights of outgoing links from nodes in group r and K_s^{in} is the sum of weights of incoming links to nodes in group s. K_s and K_r play the role of the occupation numbers n_r and n_s in (3.20). Note that this form of p_{ij} does not ensure $p_{ij} < 1$ for all i, j but this little inconsistency does not seem to have a large impact in practice and in particular for sparse networks.

Note that it is possible to also include degree–degree correlations or any other form of prior knowledge about p_{ij} at this point. For instance, we may first compute a hidden variable model [7] to reproduce the observed degree distribution including their correlations and link reciprocity and use the hence computed values of p_{ij} as random null model. Though in principle p_{ij} could take any form, for an efficient optimization it is convenient to have a form which factorizes because then, the expectation values (3.20) and (3.22) can be calculated conveniently.

3.2.5 Cohesion and Adhesion

From the above considerations and to simplify further developments, the concepts of "cohesion" and "adhesion" are introduced. The coefficient of adhesion between groups r and s is defined as

$$a_{rs} = m_{rs} - \gamma [m_{rs}]_{p_{ij}}. \tag{3.23}$$

For $r = s$, we call $c_{ss} = a_{ss}$ the coefficient of "cohesion". Two groups of nodes have a positive coefficient of adhesion, if they are connected by edges bearing more weight than expected from p_{ij}. We hence call a group cohesive, if its nodes are connected by edges bearing more weight than expected from p_{ij}. This allows for a shorthand form of (3.15) as $Q = \frac{1}{2} \sum_{rs} |a_{rs}|$ and we see that the block model $\mathbf{B}$ has entries of one where $a_{rs} > 0$. Remember that a_{rs} depends on the global parameter γ and the assumed penalty function p_{ij}. For $\gamma = 1$ and the model $p_{ij} = \frac{k_i^{out} k_j^{in}}{M}$ one finds

$$\sum_{rs} a_{rs} = \sum_r a_{rs} = \sum_s a_{rs} = 0. \tag{3.24}$$

This means that when $\mathbf{B}$ is assigned from (3.15) there exists at least one entry of one and at least one entry of zero in every row and column of $\mathbf{B}$ (provided that the network is not complete or zero).

3.2.6 Optimizing the Quality Function

After having studied some properties of the configurations and image graphs that optimize (3.13), (3.14) or (3.15), let us now turn to the problem of actually finding these configurations. Though any optimization scheme that can deal with combinatorial optimization problems may be implemented [8, 9], the use of simulated annealing [10] for a Potts model [11] is shown, because it yields high-quality results, is very general in its application and very simple to program. We interpret our quality function Ω to be maximized as the negative of a Hamiltonian to be minimized, i.e., we write $\mathcal{H}(\{\sigma\}) = -\Omega$. The single site heat bath update rule at temperature $T = 1/\beta$ then reads as follows:

$$p(\sigma_i = \alpha) = \frac{\exp\left(-\beta\mathcal{H}(\{\sigma_{j\neq i}, \sigma_i = \alpha\})\right)}{\sum_{s=1}^{q}\exp\left(-\beta\mathcal{H}(\{\sigma_{j\neq i}, \sigma_i = s\})\right)}. \qquad (3.25)$$

That is, the probability of node i being in group α is proportional to the exponential of the energy (negative quality) of the entire system with all other nodes $j \neq i$ fixed and node i in state α. Since this is costly to evaluate, one pretends to know the energy of the system with node i in some arbitrarily chosen group ϕ, which is denoted by $\mathcal{H}_\phi$. Then one can calculate the energy of the system with i in group α as $\mathcal{H}_\phi + \Delta\mathcal{H}(\sigma_i = \phi \to \alpha)$. The energy $\mathcal{H}_\phi$ then factors out in (3.25) and one is left with

$$p(\sigma_i = \alpha) = \frac{\exp\left\{-\beta\Delta\mathcal{H}(\sigma_i = \phi \to \alpha)\right\}}{\sum_{s=1}^{q}\exp\left\{-\beta\Delta\mathcal{H}(\sigma_i = \phi \to s)\right\}}. \qquad (3.26)$$

Suppose we are trying to fit a network to a given image graph, i.e., $\mathbf{B}$ is given. Then the change in energy $\Delta\mathcal{H}(\sigma_i = \phi \to \alpha)$ is easily calculated from the change in quality according to (3.13):

$$\Delta\mathcal{H}(\sigma_i = \phi \to \alpha) = \sum_s (B_{\phi s} - B_{\alpha s})(k_{i \to s}^{out} - \gamma[k_{i \to s}^{out}])$$

$$+ \sum_r (B_{r\phi} - B_{r\alpha})(k_{r \to i}^{in} - \gamma[k_{r \to i}^{in}]) \qquad (3.27)$$

$$= \sum_s (B_{\phi s} - B_{\alpha s})a_{is} + \sum_r (B_{r\phi} - B_{r\alpha})a_{ri}. \qquad (3.28)$$

Here $k_{i \to s}^{out} = \sum_j w_{ij}A_{ij}\delta_{\sigma_j,s}$ denotes the number (weight) of outgoing links node i has to nodes in role s and $[k_{i \to s}^{out}] = \sum_j p_{ij}\delta_{\sigma_j,s}$ denotes the respective expectation value. Further, $k_{r \to i}^{in} = \sum_j w_{ji}A_{ji}\delta_{\sigma_j,r}$ denotes the number (weight) of incoming links node i has from nodes in role r and $[k_{r \to i}^{out}] = \sum_j p_{ji}\delta_{\sigma_j,r}$ denotes the expectation value. By a_{is} we thus understand the adhesion of node i to all nodes in group s. For undirected networks, the two contributions of incoming and outgoing links are of course equal. Hence, our single site updating scheme needs to assess the k_i neighbors of node i and then to determine which of the q roles is best for this node, which takes $\mathcal{O}(q^2)$ operations. Thus, a local update needs $\mathcal{O}(\langle k \rangle + q^2)$ operations and can be implemented efficiently on sparse graphs as long as the number of roles is much smaller than the number of nodes in the network. Naturally, the optimal assignment of roles to nodes is characterized by $\Delta\Omega(\sigma_i = \alpha \to \phi) \leq 0$, i.e., every node assumes its best-fitting role, provided all other nodes do not change.

The optimization of (3.15) without a given image graph $\mathbf{B}$ is different, but its computational complexity remains the same. The change in energy is then calculated from the change in quality according to (3.15):

$$\Delta \mathcal{H}(\sigma_i = \phi \rightarrow \alpha) = \sum_s^q |a_{\phi s}| + |a_{\alpha s}| - |a_{\phi-i,s}| - |a_{\alpha+i,s}|$$

$$+ \sum_r^q |a_{r\phi}| + |a_{r\alpha}| - |a_{t,\phi-i}| - |a_{t,\alpha+i}|. \quad (3.29)$$

Here, a_{rs} is the coefficient of adhesion already defined in (3.23). The subscript $\phi - i$ denotes all nodes in group ϕ except i and $\alpha + i$ denotes all nodes in group α plus i. For the two models of link distribution introduced, these coefficients of adhesion are efficiently calculated. For $p_{ij} = p$ we find

$$a_{rs} = m_{rs} - \gamma p n_r n_s, \quad (3.30)$$

$$a_{r\pm i,s} = m_{rs} + k_{i \rightarrow s}^{out} - \gamma p(n_r \pm 1)n_s, \quad (3.31)$$

$$a_{r,s\pm i} = m_{rs} - k_{r \rightarrow i}^{in} - \gamma p n_r (n_s \pm 1). \quad (3.32)$$

And for $p_{ij} = k_i^{out} k_j^{in}/M$ we can write

$$a_{rs} = m_{rs} - \gamma K_r^{out} K_s^{in}, \quad (3.33)$$

$$a_{r\pm i,s} = m_{rs} \pm k_{i \rightarrow s}^{out} - \gamma(K_r^{out} \pm k_i^{out})K_s^{in}, \quad (3.34)$$

$$a_{r,s\pm i} = m_{rs} \pm k_{r \rightarrow i}^{in} - \gamma K_r^{out}(K_s^{in} \pm k_i^{in}). \quad (3.35)$$

At this point it becomes clear why a form of p_{ij} which factorizes on the level of individual nodes is so convenient. We see from (3.30–3.35) that the expectation values for the number of links (resp. link weights) also factorize and one needs to keep track of only q global values $K_s^{in/out}$ or n_r in order to calculate the single site update probabilities.

3.3 Conclusion

In this chapter we have derived a new quality function for block models in complex networks. The fundamental observation was that in the approximation of sparse networks by block structures it makes sense to give more weight to matching the sparse edges than to matching the abundant missing edges. This leads to a function which bases the detection of patterns on the detection of maximal deviations from expected behavior according to a random null model in accordance with our initial definition of a pattern as everything which is not random. The quality function we defined is formally equivalent to a model of magnetic materials, the so-called Potts model. We will exploit this similarity greatly in the next chapters. In the following, we will focus on diagonal block models, i.e., modular structures or cohesive subgroups which form most likely the most important sub-class of block models and have received the greatest attention during recent years.

References

1. O. Martin, R. Monasson, and R. Zecchina. Statistical mechanics methods and phase transitions in optimization problems. *Theoretical Computer Science*, 256:3–67, 2001.
2. Y. Fu and P. W. Anderson. Application of statistical mechanics to NP-complete problems in combinatorial optimisation. *Journal of Physics A: Mathematical and General*, 19:1605–1620, 1986.
3. M. Weigt and A. K. Hartman. Number of guards needed by a museum: A phase transition in vertex covering of random graphs. *Physical Review Letters*, 84(26):6118, 2000.
4. R. Monasson and R. Zecchina. Entropy of the k-satisfiability problem. *Physical Review Letters*, 76:3881, 1996.
5. V. Černy. Thermodynamical approach to the travelling salesman problem: An efficient simulation algorithm. *Journal of Optimization Theory and Applications*, 45:41, 1985.
6. F. Y. Wu. The Potts model. *Reviews of Modern Physics*, 54(1):235–368, 1982.
7. D. Garlaschelli and M. I. Loffredo. Maximum likelihood: extracting unbiased information from complex networks. *Physical Review E*, 7: 78(1):015101, 2008.
8. A. K. Hartmann and H. Rieger, editors. *Optimization Algorithms in Physics*. Wiley-Vch, Weinheim, 2004.
9. A. K. Hartmann and H. Rieger, editors. *New Optimization Algorithms in Physics*. Wiley-Vch, Weinheim, 2004.
10. S. Kirkpatrick, C.D. Gelatt Jr., and M.P. Vecchi. Optimization by simulated annealing. *Science*, 220:671–680, 1983.
11. M. E. J. Newman and G. T. Barkema. *Monte Carlo Methods in Statistical Physics*. Clarendon Press, Oxford, 1999.

Diagonal Block Models as Cohesive Groups

The importance of cohesive subgroups in networks was already discussed in Chap. 2. In this chapter, we will discuss communities or cohesive subgroups as a special class of block models in the framework introduced in the last chapter.

4.1 Equivalence with Newman–Girvan Modularity and Spin Glass Energy

In Chap. 2, a function to assess the quality of a community structure was introduced, the so-called modularity Q defined by Newman and Girvan [1]. It can be shown that Q is a special case of the universal *ansatz* presented in Sect. 3.2. Newman and Girvan's modularity measure is written as [1]:

$$Q = \sum_s e_{ss} - a_s b_s, \text{ with } a_s = \sum_r e_{rs} \text{ and } b_s = \sum_r e_{sr}. \qquad (4.1)$$

Here, e_{rs} is the fraction of links that fall between nodes in group r and s, i.e., the probability that a randomly drawn link connects a node in group r to one in group s. The probability that a link has at least one end in group s is expressed by a_s. From this, one expects a fraction of $a_s b_s$ links to connect nodes in group s among themselves. Newman's modularity measure hence compares the actual link density in a community with an expectation value based on the row and column sums of the matrix e_{rs}. One can write this modularity in a slightly different way following [2]:

$$e_{ss} = \frac{1}{M} \sum_{ij} A_{ij} \delta(\sigma_i, s) \delta(\sigma_j, s)$$

$$a_s b_s = \left(\frac{1}{M} \sum_i k_i^{out} \delta(\sigma_i, s) \right) \left(\frac{1}{M} \sum_i k_i^{in} \delta(\sigma_i, s) \right)$$

Reichardt, J.: *Diagonal Block Models as Cohesive Groups*. Lect. Notes Phys. **766**, 45–68 (2009)
DOI 10.1007/978-3-540-87833-9_4

$$Q = \frac{1}{M} \sum_{ij} \left(A_{ij} - \frac{k_i^{out} k_j^{in}}{M} \right) \delta(\sigma_i, \sigma_j). \qquad (4.2)$$

This already resembles (3.13) when p_{ij} takes the form $k_i^{out} k_j^{in}/M$, $\gamma = 1$ and $B(r, s) = \delta(r, s)$. To maximize the modularity of a community structure is hence equivalent to finding the optimal matching of a network to an image graph $\mathbf{B}$ which consists of isolated nodes with self-links only. It is worth noting that the computational complexity of the single node update rule of (3.28) becomes $\mathcal{O}(\langle k \rangle + q)$ because of the diagonal structure of $\mathbf{B}$. This allows the use of very large, though diagonal, image graphs or a partition into very many cohesive subgroups.

Thus far, it was shown that the configuration of group indices which maximizes (4.2) can be interpreted as *the* community structure of a network. Formally, (4.2) is equivalent to the negative of a Hamiltonian of a Potts spin glass with couplings between every pair of nodes. The couplings are strongly ferromagnetic along the links of the graph and weakly anti-ferromagnetic between nodes which are not linked. The lower the energy of this spin glass, the "better", i.e., more modular, the community structure. The best assignment into communities is hence found in the configuration with minimal energy, i.e., in the ground state of

$$\mathcal{H} = -\sum_{ij} \left(A_{ij} - \gamma \frac{k_i^{out} k_j^{in}}{M} \right) \delta(\sigma_i, \sigma_j). \qquad (4.3)$$

Note that we have dropped the normalizing factor $1/M$ to make this Hamiltonian extensive. The fact that the modularity shows a formal equivalence with a model of a spin glass will allow us to derive numerous insights into the behavior of this quality function. In particular, it will allow us to use the full machinery of statistical mechanics to derive expectation values for the modularity of different classes of random networks which are indispensable for the evaluation of the statistical significance of the findings of our block modeling procedure. For convenience, from this point forward, we will discuss everything in terms of minimal energy (4.3) instead of maximum quality fit (3.13).

4.1.1 Properties of the Ground State

From the fact that the ground state is a configuration which is a minimum in the configuration space, one can derive a number of properties of the communities that apply to any local minimum of the Hamiltonian in the configuration space. If one takes these properties as *defining* properties of what a community is, one then finds valid alternative community structures also in the local minima of the Hamiltonian. The energies of these local minima will then allow us to compare these community structures. It may be that alternative but almost equally "good" community structures exist. Before proceeding to investigate the properties of spin configurations that represent local minima of the Hamiltonian, a few properties of (4.3) as such shall be discussed:

First, note that for $\gamma = 1$ (4.3) evaluates to zero in case of assigning all nodes into the same spin state due to the normalization constraint on p_{ij}, i.e., $\sum_{ij} p_{ij} = \sum_{ij} A_{ij} = M$, independent of the graph. Second, for a complete graph, any spin configuration yields the same zero energy at $\gamma = 1$. Third, for a graph without edges, e.g., only a set of nodes, any spin configuration gives zero energy independent of γ. Fourth, the expectation value of (4.3) for a random assignment of spins at $\gamma = 1$ is zero. These considerations provide an intuitive feeling for the fact that the lower the energy the better the fit of the diagonal block model to the network and that the choice of $\gamma = 1$ will result in what could be called "natural partitioning" of the graph into modules.

Let us consider the case of undirected networks which is most often found in applications. Then, the adjacency matrix of the network is symmetric and we have $k_i^{in} = k_i^{out}$ and thus the coefficients of adhesion are also symmetric, i.e., $a_{rs} = a_{sr}$. According to (3.28) the change in energy to move a group of nodes n_1 from group s to spin state r is

$$\Delta \mathcal{H} = a_{1,s\backslash 1} - a_{1r}. \tag{4.4}$$

Here $a_{1,s\backslash 1}$ is the adhesion of n_1 with its complement in group s and a_{1r} is the adhesion of n_1 with n_r. It is clear that if one moves n_1 to a previously unpopulated spin state, then $\Delta \mathcal{H} = a_{1,s\backslash 1}$. This move corresponds to dividing group n_s. Furthermore, if $n_1 = n_s$, one has $\Delta \mathcal{H} = -a_{sr}$, which corresponds to joining groups n_s and n_r. A spin configuration can only be a local minimum of the Hamiltonian if a move of this type does not lead to a lower energy. It is clear that some moves may not change the energy and are hence called neutral moves. In cases of equality $a_{1,s\backslash 1} = a_{1,r}$ and n_r being a community itself, communities n_s and n_r are said to have an overlap of the nodes in n_1.

For a community defined as a group of nodes with the same spin state in a spin configuration that makes the Hamiltonian (4.3) minimal, one then has the following properties:

(i) Every proper subset of a community has a maximum coefficient of adhesion with its complement in the community compared to the coefficient of adhesion with any other community ($a_{1,s\backslash 1} = \max$).
(ii) The coefficient of cohesion is non-negative for all communities ($c_{ss} \geq 0$).
(iii) The coefficient of adhesion between any two communities is non-positive ($a_{rs} \leq 0$).

The first property is proven by contradiction from the fact that one is dealing with a spin configuration that makes the Hamiltonian minimal. One also observes immediately that every proper subset n_1 of a community n_s must have a non-negative adhesion with its complement $n_{s\backslash 1}$ in the community. In particular this is true for every single node l in n_s ($a_{l,s\backslash l} \geq 0$). Then one can write $\sum_{l \in n_s} a_{l,s \setminus l} \geq 0$. Since $\sum_{l \in n_s} m_{l,s \setminus l} = 2m_{ss}$ and $\sum_{l \in n_s} [m_{l,s \setminus l}]p_{ij} = 2[m_{ss}]p_{ij}$, this implies $c_{ss} \geq 0$ for all communities s and proves the second property. The third property is proven by contradiction again. Please note that for $\gamma = 1$ and $p_{ij} = k_i k_j / M$, no community is formed of a single node due to condition (3.24). The last two properties can be summarized in the following inequality which provides an intuition about the significance of the parameter γ:

$$c_{ss} \geq 0 \geq a_{rs} \quad \forall r \neq s. \tag{4.5}$$

Assuming a constant link probability, one can rewrite this inequality in order to relate the inner link density of a community and the outer link density between communities with an average link density:

$$\frac{2m_{ss}}{n_s n_s} \geq \gamma p \geq \frac{m_{rs}}{n_r n_s} \quad \forall r \neq s. \tag{4.6}$$

Note that γp can be interpreted as a threshold between inner and outer link density under the assumption of a constant link probability.

Apart from giving an interpretation of the (local) minima of the Hamiltonian, the above properties also give a *definition* of what a community is, alternative to that of a set of nodes of equal spin value in a configuration that represents a minimum of the Hamiltonian. When speaking of the community structure of a network, one generally refers to that obtained at lowest energy, i.e., in the ground state. One can also use the term "community" denoting a subset of nodes that has all of the above properties. Note that this definition of community adapts itself naturally to different classes of networks, since a model p_{ij} is included in the definition of adhesion and cohesion. Since the assignment of nodes into communities changes with the value of γ, the notion "community at level γ" shall be adopted, in order to characterize possible hierarchies in the community structure.

4.1.2 Simple Divisive and Agglomerative Approaches to Modularity Maximization

The equivalence of modularity with a spin glass energy shows that the problem of maximizing modularity falls into the class of NP-hard optimization problems [3]. For these problems, it is believed that no algorithm exists that is able to produce an optimal solution in a time that grows only polynomial with the size of the problem instance. However, heuristics such as simulated annealing exist, which are able to find possibly very good solutions. In this section, we will discuss an often used approach to clustering, namely hierarchical agglomerative and divisive algorithms and investigate whether they too are good heuristics for finding partitions of maximum modularity.

A number of community detection algorithms presented in Chap. 2 have followed recursive approaches and lead to hierarchical community structures. Hierarchical clustering techniques can be dichotomized into divisive and agglomerative approaches [4]. It will be shown how a simple recursive divisive approach and an agglomerative approach may be implemented and where they fail.

In the present framework, a hierarchical divisive algorithm would mean to construct the ground state of the q-state Potts model by recursively partitioning the network into two parts according to the ground state of a 2-state Potts or Ising system. This procedure would be computationally simple and result directly in a hierarchy of clusters due to the recursion of the procedure on the parts until the total energy cannot be lowered anymore. Such a procedure would be justified, if the ground state of the q-state Potts Hamiltonian and the repeated application of the Ising system cut the network along the same edges. Let us derive a condition under which this could be ensured.

In order for this recursive approach to work, one must ensure that the ground state of the 2-state Hamiltonian never cuts though a community as defined by the q-state Hamiltonian. Assume a network made of three communities n_1, n_2 and n_3 as defined by the ground state of the q-state Hamiltonian. For the bi-partitioning, one now has two possible scenarios. Without loss of generality, the cut is made either between n_2 and $n_1 + n_3$ or between n_1, n_2 and $n_3 = n_a + n_b$, parting the network into $n_1 + n_a$ and $n_2 + n_b$. Since the former situation should be energetically lower for the recursive algorithm to work, one arrives at the condition that

$$m_{ab} - [m_{ab}]_{p_{ij}} + m_{1b} - [m_{1b}]_{p_{ij}} > m_{2b} - [m_{2b}]_{p_{ij}}, \qquad (4.7)$$

which must be valid for all subgroups n_a and n_b of community n_3. Since n_3 is a community, it is further known that $m_{ab} - [m_{ab}]_{p_{ij}} > m_{1b} - [m_{1b}]_{p_{ij}}$ and $m_{ab} - [m_{ab}]_{p_{ij}} > m_{2b} - [m_{2b}]_{p_{ij}}$. Though $m_{ab} - [m_{ab}]_{p_{ij}} > 0$, since n_3 is a community, $m_{1b} - [m_{1b}]_{p_{ij}} < 0$ and $m_{2b} - [m_{2b}]_{p_{ij}} < 0$ for the same reason and hence condition (4.7) is not generally satisfied. Figure 4.1 illustrates a counterexample.

Assuming $p_{ij} = p$, part (a) of the figure shows the ground state of the system when using only two spin states. Part (b) of Fig. 4.1 shows the ground state of the system without constraints on the number of spin states, resulting in a configuration of three communities. The bi-partitioning approach would have cut through one of the communities in the network. Recursive bi-partitionings cannot generally lead to an optimal assignment of spins that maximizes the modularity.

In [5] Newman et al. have introduced a fast greedy strategy for modularity maximization. It effectively corresponds to a simple nearest neighbor agglomerative clustering of the network where the adhesion coefficient a_{rs} is used as a similarity measure. The algorithm initially assigns different spin states to every node and then proceeds by grouping those nodes together that have the highest coefficient of adhesion. As Fig. 4.2 shows, this approach fails, if the links between two communities connect nodes of low degree. The network consists of 14 nodes and 37 links. It is clearly seen that in the ground state,

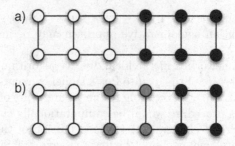

Fig. 4.1. Illustration of the problem of recursive bi-partitioning. The ground state of the Hamiltonian with only two possible spin states, as shown in (**a**), would cut through one of the communities that are found when allowing three spin states as shown in (**b**).

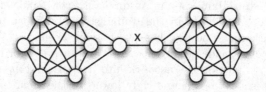

Fig. 4.2. Example network for which an agglomerative approach of grouping together nodes of maximal adhesion will fail. Starting from an assignment of different spin states to every node, the largest adhesion is found for the nodes connected by edge x, which are grouped together first by the agglomerative procedure. However, it is clearly seen that x should lie between different groups.

the network consists of two communities and edge x lies between them. However, when initially assigning different spin states to all nodes, the adhesion a between the nodes connected by x is largest: $a = 1 - 16/2M$, since the product of degrees at this edge is lowest. Therefore, the agglomerative procedure described is misled into grouping together the nodes connected by x already in the very first step. Furthermore, it is clear that in a network, where all nodes have the same degree initially, all edges connect nodes of the same coefficient of adhesion. In this case, it cannot be decided which nodes to group together in the first step of the algorithm at all. It was shown by Newman [5] that the approach does deliver good results in benchmarks using computer-generated test networks as introduced in Sect. 2.4. The success of this approach depends of course on whether or not the misleading situations have a strong effect on the final outcome of the clustering. In the example shown, after grouping together the nodes at the end points of x, the algorithm will proceed to further add nodes from only one of the two communities linked by x. Hence, the initial mistake persists, but does not completely destroy the result of the clustering.

4.1.3 Finding the Community Around a Given Node

Often, it is desirable not to find all communities in a network, but to find only the community to which a particular node belongs. This may be especially useful if the network is very large and detecting all communities may be time consuming. In the framework presented here, one can do this using a fast, greedy algorithm. Starting from the node one is interested in, node j, one successively adds nodes with positive adhesion to the group and continues as long as the adhesion of the community one is forming with the rest of the network is decreasing. Adding a node i from the rest of the network r to the community s around the start node j, the adhesion between s and r changes by

$$\Delta a_{sr}(i \to s) = a_{ir} - a_{is}. \tag{4.8}$$

For $p_{ij} = p$, this can be written as

$$\Delta a_{sr}(i \to s) = k_{ir} - k_{is} - \gamma p(n_r - 1 - n_s), \tag{4.9}$$

where $n_r = N - n_s$ is the number of nodes in the rest of the network and n_s the number of nodes in the community. For $p_{ij} = k_i k_j / M$, the change in adhesion reads

$$\Delta a_{sr}(i \to s) = k_{ir} - k_{is} - \frac{\gamma}{M} k_i (K_r - k_i - K_s). \tag{4.10}$$

Here, K_r and K_s are the sums of degrees of the rest of the network r and the community under study s, respectively, and k_i is the degree of node i to be moved from r to s, which has $k_{i \to s}$ links connecting it with s and $k_{i \to r}$ links connecting it with the rest of the network. It is understood that only when the adhesion of i with s is larger than with r, the total adhesion of s with r decreases. Equivalent expressions can be found for removing a node i from the community s and rejoining it with r. The generalization to directed networks is then straightforward. For $\gamma = 1$ and $p_{ij} = k_i k_j / M$, one has $a_{is} + a_{ir} + 2c_{ii} = 0$, and $c_{ii} < 0$ by definition for networks without self-links and close to zero for all practical cases. Then, a_{is} and a_{ir} are either both positive and very small or have opposite signs.

Choosing the node that gives the smallest Δa_{rs} will then result in adding a node with positive coefficient of adhesion to s. It is easy to see that this ensures a positive coefficient of cohesion in the set of nodes around j.

4.2 Comparison with Other Definitions of Communities

In Sect. 4.1.1 the term community was defined as a set of nodes having properties (*i*) through (*iii*). Compared with the many definitions of community in the sociological literature [6], this definition is most similar to that of an "LS set". Recall, LS set is a set of nodes S in a network such that each of its proper subsets has more links to its complement in S than to the rest of the network [7]. Note, however, that the problem in the definition of an LS set mentioned in Sect. 2.3.1 does not occur.

Previously, Radicchi et al. [8] had given a definition of community "in a strong sense" as a set of nodes V with the condition $k_i^{in} > k_i^{out}, \forall i \in V$, i.e., every node in the group has more links to other members of the group than to the rest of the network. In the same manner, they define a community in a "weak sense" as a set of nodes V for which $\sum_{i \in V} k_i^{in} > \sum_{i \in V} k_i^{out}$, i.e., the total number of internal links is larger than half of the number of the external links, since the sum of k_i^{in} is twice the number of internal edges. The similarity with properties (1) and (2) of the new definition is evident, but instead of comparing absolute numbers for single nodes, the new definition compares absolute numbers to expectation values for these quantities in the form of the coefficients of cohesion and adhesion not only for single nodes but also for sets of nodes. As already discussed in Sect. 2.3.2, one of the consequences of Radicchi et al.'s definitions is that every union of two communities is also a community. This leads to the strange situation that a community in the "strong" or "weak" sense can also be an ensemble of disjoint groups of nodes. This paradox may only be resolved if one assumes a priori that there exists a hierarchy of communities. The following considerations and examples will show that hierarchies in community structures are possible, but cannot be taken for granted. The representation of community structures by dendrograms, therefore, cannot always capture the true community structure and hence all hierarchical community detection algorithms should be used with caution.

Another definition of communities is that given by Palla et al. [9,10], with a community defined as a set of nodes that can be reached through a clique percolation process. Apart from the issues already identified in Sects. 2.3.1 and 2.4.3, let us stress again the difference in the definition of overlap. The k-clique percolation process implies a nested hierarchy in the sense that a k+1-clique is always entirely contained in a k-clique, though it does also allow for overlap in the sense that nodes may be part of more than one k-clique. The overlapping nodes themselves, however, can never form a k-clique themselves. That is, the overlapping nodes can never be a community of their own. As will be shown below, this situation is possible and is indeed encountered in real world networks as well.

4.2.1 Hierarchy and Overlap of Community Assignments

Even though hierarchical community structures cannot be taken for granted and hence should not be enforced by using hierarchical community detection algorithms, they still form an important organizational principle in networks which shall be investigated directly from the adjacency matrix. When ordering the rows and columns according to the assignment of nodes into communities, the link density in the adjacency matrix is directly transformed into point density and hence into gray levels. Since the inner link density of a community is higher than the external, one can distinguish communities as square blocks of darker gray. Different orderings may be combined into a consensus ordering. That is, starting from a super-ordering given, the nodes within each community are reordered according to a second given sub-ordering, i.e., one only changes the internal order of the nodes within communities of the super-ordering. This leads to the formation of new blocks of those nodes that are assigned together in one community in both orderings. One can then repeat the procedure to obtain further iterative consensus orderings.

First, an example of a completely hierarchical network is given very similar to that used in Ref. [11]. Here, hierarchy implies that all communities found at a value of $\gamma_2 > \gamma_1$ are proper sub-communities of the communities found at γ_1. In the example, a network made of four large communities of 128 nodes each was constructed. Each of these nodes has an average of 7.5 links to the 127 other members of their community and 5 links to the remaining 384 nodes in the network. Each of these 4 communities is composed of 4 sub-communities of 32 nodes each. Each node has an additional 10 links to the 31 other nodes in its sub-community. Figure 4.3 shows the adjacency matrix of this network in different orderings.

At $\gamma = 1$, the ground state is composed of the four large communities as shown in the left part of Fig. 4.3. Increasing γ above a certain threshold makes assigning different spin states to the 16 sub-communities the ground state configuration. The middle part of Fig. 4.3 shows an ordering obtained with a value of $\gamma = 2.2$. One can see that some of the these sub-communities are more densely connected among each other. Imposing the latter ordering on top of the ordering obtained at $\gamma = 1$ then allows to display the full community structure and hierarchy of the network as shown in the right part of Fig. 4.3. Note that a recursive approach applying the community detection

Fig. 4.3. Example of an adjacency matrix for a perfectly hierarchical network. The network consists of four communities, each of which is composed of four subcommunities. Using $\gamma = 1$, the four main communities (*left*) are found. With $\gamma = 2.2$, one finds the 16 sub-communities (*middle*). Link density variations in the off diagonal parts of the adjacency matrix already hint at a hierarchy. The consensus ordering (*right*) shows that each of the larger communities is indeed composed of four sub-communities each.

Fig. 4.4. Example of an adjacency matrix for an only partially hierarchical network with overlapping community structure. The network consists of two large communities A and B, each of which contains a sub-community a and b, which are densely linked with each other. Using $\gamma = 0.5$, one finds the two large communities (*left*). With a larger $\gamma = 1$, the two small sub-communities a and b are grouped together. The consensus ordering (*right*) shows that most of the links which join A and B in fact lie between a and b.

algorithm to separate subgroups was *not* used. Instead, two *independent* orderings were obtained which are only compatible with each other, because the network has a hierarchical structure of dense communities composed of denser subcommunities.

In contrast to this situation, Fig. 4.4 shows an example of a network that is only partially hierarchical. The network consists of 2 large communities A and B containing 512 nodes, which have on average of 12 internal links per node. Within A and B, a subgroup of 128 nodes exists, which is denoted by a and b, respectively. Every node within this subgroup has 6 of its 12 intra-community links to the 127 other members of this subgroup. The two subgroups a and b have on average three links per node with each other. Additionally, every node has two links with randomly chosen nodes from the network. From Fig. 4.4, one observes that the two large communities are found using $\gamma = 0.5$. Maximum modularity, however, is reached at $\gamma = 1$ when a and b are joined into a separate community.

Only when using the consensus of the ordering obtained at $\gamma = 0.5$ and $\gamma = 1$, one can understand the full community structure with a and b being subgroups that are responsible for the majority of links between A and B. It is understood, that this situation cannot be interpreted as a hierarchy, even though a and b are cohesive subgroups in A and B, respectively. Here, the nodes responsible for the overlap form a community of their own at a particular value of γ.

From these examples, it is clear that the above link structures would be much better fitted with a non-diagonal block model. However, the reduced computational cost of fitting a diagonal block model may sometimes make the procedure outlined above useful.

One cannot generally assume that a community structure of a network is uniquely defined. There may exist several but very different partitions that all have a very high value of modularity. Palla et al. [9, 10] have introduced an algorithm to detect overlapping communities by clique percolation and Gfeller et al. have introduced the notion of nodes lying "between clusters" [12]. This uncertainty in the definition of the borders of a community is expected from the spin glass nature of the Hamiltonian, where generally many energy minima may exist that are comparably deep corresponding to comparably good assignments of nodes into communities. Additionally, the (local) minima of the Hamiltonian may be degenerate. The overlap of communities is linked to degeneracy of the minima of the Hamiltonian. Since the degeneracy can arise in several ways, one has to differentiate between two different types of overlap: overlap of community structure and overlap of communities.

It was already shown that it is undecidable whether a group of nodes n_t should be a member of community n_s or n_r, if the coefficients of adhesion are equal for both of these communities. Formally, one finds $a_{t,s\setminus t} = a_{tr}$. In this situation, one speaks of overlapping communities n_s and n_r with overlap n_t, since the number of communities in the network is not affected by this type of degeneracy. Nodes that do not form part of overlaps will always be grouped together and can be seen as the non-overlapping cores of communities.

On the other hand, it may be undecidable, if two groups of nodes should be grouped together or apart, if the coefficient of adhesion between them is zero, i.e., there exist as many edges between them as expected from the model p_{ij}. Similarly, it may be undecidable, if a group of nodes should form its own community or be divided and the parts joined with different communities, if this can be done without increasing the energy. In these situations, the number of communities in the ground state is not well defined and one cannot speak of overlapping communities, since communities do not share nodes in the degenerate realizations. Hence, such a situation shall be referred to as overlapping community structures.

The overlap is best represented in a symmetric $N \times N$ "co-appearance" matrix, in which the entry i, j denotes if or how often nodes i and j were grouped together. This also allows the combined representation of the overlap of many different community structures necessary for the investigation of degenerate ground states. In addition, this type of representation allows the comparison of community structures obtained from a parameter variation of γ also for large networks and hence to study possible hierarchies and the stability of community structures for different values of γ when possible degenerate ground states can only be sampled in a stochastic manner.

It was already stressed that properties (i) through (iii) are also valid for any local minimum of the energy landscape defined by the Hamiltonian and the graph. They only imply that one cannot jump over energy barriers and move into deeper minima using the suggested move set. It may therefore be interesting to study also the local minima and compare them to the ground state. Local minima may be sampled by running greedy optimization algorithms using random initial conditions. For correlated energy landscapes, it is known that deeper local minima have larger basins of attraction in the configuration space [13]. The Hamiltonian (4.3) induces such a correlated energy landscape on the graph, since the total energy is not drastically affected by single spin changes. Therefore, one expects that the deep local minima will be sampled with higher frequency and that pairs of nodes that are grouped together in deep minima will have larger entries in the co-appearance matrix. A number of examples of co-appearance matrices sampling local energy minima at different values of γ will be given later.

The use of co-appearance matrices derived from sampling the local minima of our quality function is of course not limited to the analysis of diagonal block models but can also be used to study the stability of role assignments for non-diagonal block models as well.

4.3 Benchmarking the Algorithm

In order to benchmark the performance of the Potts model approach to community detection, it is applied to computer-generated test networks. Networks with communities of equal and different size were constructed. Those with equal size had 128 nodes, grouped into 4 communities of size 32. Those with differently sized communities had 320 nodes, grouped into 4 communities of size 32, 64, 96 and 128. In both types of networks, each node has an average degree of $\langle k \rangle = 16$. The average number of links to members of the same community $\langle k_{in} \rangle$ and to members of different communities $\langle k_{out} \rangle$ is then varied, but always ensuring $\langle k_{in} \rangle + \langle k_{out} \rangle = \langle k \rangle$. Hence, decreasing k_{in} renders the problem of community detection more difficult.

Recovering a known community structure, any algorithm has to fulfill two criteria: it has to group nodes in the same community which belong together by design and it has to group nodes apart which belong to different communities by design. The first criterion is called "sensitivity" and measures the percentage of *pairs of nodes* which are correctly grouped together. The second criterion is called "specificity" and measures the percentage of pairs of nodes which are correctly grouped apart.

Because of the Poisson nature of the degree distribution, a connection model of $p_{ij} = p$ was used. Figure 4.5 shows the result of this experiment in comparison with the results obtained from the algorithm of Girvan and Newman [14]. Clearly, both algorithms show high sensitivity and high specificity. However, the Potts model outperforms the GN algorithm on both types of networks in both sensitivity and specificity. When relaxing the Potts model Hamiltonian from random initial conditions at zero temperature, performance decreases, but is still as good as that of the GN algorithm.

An important aspect is the dependence of the sensitivity (specificity) of the algorithm on the number of allowed spin states q. Figure 4.6 shows that as long as $q \geq 4$, i.e., the actual number of communities in the network, the value of q is irrelevant. This result is also independent of the strength of the

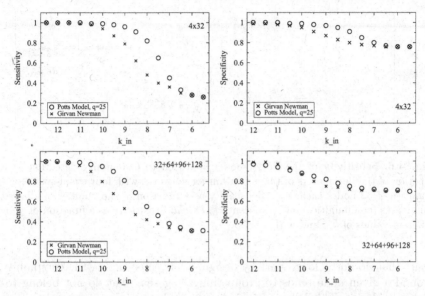

Fig. 4.5. Benchmarking the Potts model approach to community detection with networks of known community structure. Sensitivity measures the percentage of *pairs of nodes* correctly identified as belonging to the same community and specificity measures the percentage of pairs of nodes correctly grouped into different communities. *Top*: 4 communities of 32 nodes each. *Bottom*: 4 communities of size 32, 64, 96 and 128 nodes.

community structure under investigation, i.e., independent of k_{in}. Furthermore, it is necessary to study the stability of results with respect to a change in γ. As Fig. 4.6 shows, the better the community structure is defined, i.e., the greater k_{in} is with respect to $\langle k \rangle$, the more stable are the results. The maxima of the curves for all values of k_{in}, however, coincide at $\gamma = 1$, i.e., at the point where the contribution of missing and existing links is equal. The same statements also apply to the specificity.

In cases where exploring the community structure starts from a single node, the definitions of sensitivity and specificity have to be changed. The percentage of nodes that are correctly identified as belonging to the community around the start node is measured as sensitivity and the percentage of nodes that are correctly identified as *not* belonging to the community around the start node as specificity.

Figure 4.7 shows the results obtained for different values of $\langle k_{in} \rangle$ at $\gamma = 1$ and using $p_{ij} = k_i k_j / M$ as model of the connection probability. Note that this approach performs rather well for a large range of $\langle k_{in} \rangle$ with good sensitivity and specificity. In contrast to the benchmarks for running the simulated annealing on the entire network as shown in Fig. 4.5, a sensitivity that is generally larger than the specificity is observed. This shows that running the simulated annealing on the entire network tends to mistakenly group things

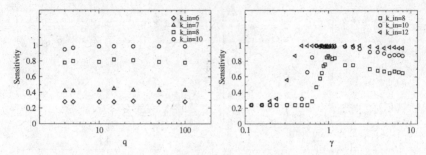

Fig. 4.6. Sensitivity of the Potts model approach to community detection as a function of the parameters of the algorithm for networks with four equal-sized communities of 32 nodes each. *Left*: Sensitivity as a function of the number of allowed spin states (communities) q for different k_{in}. *Right*: Sensitivity as a function of γ for different values of k_{in} and with $q = 25$.

apart that do not belong apart by design, while constructing the community around a given node tends to group things together that do not belong together by design. This behavior is understandable, since working on the entire network amounts to effectively implementing a divisive method, while starting from a single node means implementing an agglomerative method.

One real world example with known community structure is the college football network from Ref. [14]. It represents the game schedule of the 2000 season, Division 1, US college football league. The nodes in the network represent the 115 teams, while the links represent 613 different games played in that season. The community structure of this network arises from the grouping into

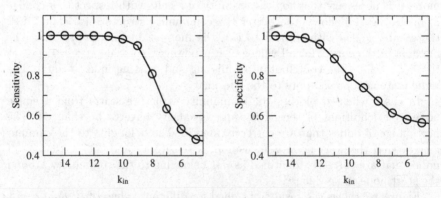

Fig. 4.7. Benchmark of the algorithm for discovering the community around a given node in networks with known community structure. Networks of 128 nodes and 4 communities were used. The average degree of the nodes was fixed to be 16, while the average number of intra-community links $\langle k_{in} \rangle$ was varied. Sensitivity measures the fraction of nodes correctly assigned to the community around the start node, while specificity measures the fraction of nodes correctly kept out of the community around the start node.

conferences of 8–12 teams, each. On average, each team has seven matches with members of its own conference and another four matches with members of different conferences. A parameter variation in γ at 10 values between $0.1 \leq \gamma \leq 1$ is performed. This allows for the estimation of the robustness of the result with respect to γ and the detection of possible hierarchies in the community structures, as low values of γ will generally lead to a less-diverse community assignment and larger communities. At each value of γ the system is relaxed 50 times from a randomly assigned initial configuration at $T = 0$ using $q = 50$. The connection model chosen was again $p_{ij} = p$.

Figure 4.8 shows the resulting 115×115 co-appearance matrix, normalized and color coded. The ordering of the matrix corresponds to the assignment of the teams into conferences according to the game schedule as indicated by the dashed lines. Apart from recovering almost exactly the known community structure, the Potts model is also able to detect inhomogeneities in the distribution of intra- and inter-conference games. For instance, one observes a large

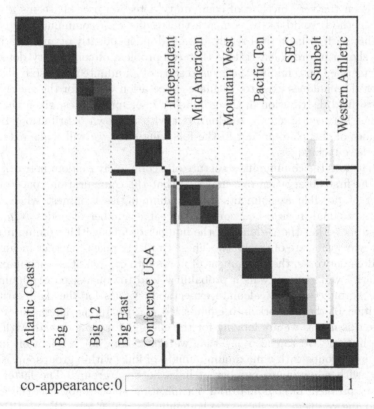

Fig. 4.8. Co-appearance matrix for the football network. A parameter variation of γ was performed with 10 values between $0.1 \leq \gamma \leq 1$. At each value, the system was relaxed 50 times from a random initial condition. The matrix ordering is taken from the assignment of teams into conferences according to the game schedule.

overlap of the Pacific Ten and Mountain West conferences and also a possible subdivision of the Mid American conference into two sub-conferences. This is due to the fact that geographically close teams are more likely to play against each other as already pointed out in Ref. [14].

4.4 Community Detection and Graph Partitioning

In order to assess the statistical significance of the community structures found with any algorithm, it is necessary to compare them with expectation values for random networks. This is of course always possible by rewiring the network randomly [15], keeping the degree distribution invariant and then running a community detection algorithm again, comparing the result to the original network. This method, however, can only give an answer to what a particular community detection algorithm may find in a random network and hence depends on the very method of community detection used. It seems a much better method would be to compare the results of a community detection algorithm with a theoretical result, obtained independently of any algorithm. It was shown in previous sections that the problem of community detection can be mapped onto finding the ground state of an infinite range spin glass. A number of techniques exist to calculate expectation values for the energy and the local field distribution in the ground state of spin glasses, given that the couplings between spins are random, but with a known distribution. In the remainder of this chapter and in the following ones, we will make extensive use of these techniques.

What does the community structure of a completely random network look like? The first observation one makes is that the configuration space or the number of possible assignments into q communities is largest when these groups are equal in size, i.e., contain the same number of nodes N/q. One can also show that the variance of the number of links within communities is largest when they are of equal size. These two facts taken together mean that in random networks, the assignment of nodes into communities with maximal modularity will with very high probability lead to equal-sized communities. Hence, we only need to calculate expectation values for the modularity of a partition of the network into equal-sized groups. In the language of spin glasses, this means we are looking for the energy of a ground state with zero magnetization. We can also say that we are looking for a partition into q equal-sized groups with a maximum number of links within groups, or, equivalently, with a minimum number of links between groups. The latter is a standard problem in combinatorial optimization and is known as the graph partitioning problem. In short, the communities which form the partition of maximum modularity in a random network will correspond to a minimum cut equipartition.

4.4.1 Expectation Values for the Modularity

The problem of graph partitioning has been studied extensively in physics literature and a number of analytical results already exist. The results presented by Fu and Anderson [16] for bi-partitioning and Kanter and Sompolinsky for q-partitioning [17] are all based on the fact that in the limit of large N, the local field distribution in infinite range systems is Gaussian and can hence be characterized by only the first two moments of the coupling distribution, the mean and the variance. The couplings used in the study of modularity are $J_{ij} = A_{ij} - \gamma p_{ij}$ which have a mean independent of the particular form of p_{ij}:

$$J_0 = (1 - \gamma)p, \tag{4.11}$$

which is zero in the case of the "natural partition" at $\gamma = 1$. The ground state of spin glasses with a coupling distribution of zero mean has always zero magnetization and hence, one must find groups of equal size [18]. This argument further backs our initial statement on the equivalence of modularity maximization in random networks and graph partitioning. The variance of the coupling distribution amounts to

$$J^2 = p - (2\gamma - \gamma^2)\langle p^2 \rangle. \tag{4.12}$$

Now one can write immediately for the modularity at $\gamma = 1$ [17]:

$$Q_q = -\frac{1}{M}\mathcal{H}_{GS} = \frac{N^{3/2}}{M}J\frac{U(q)}{q}, \tag{4.13}$$

where $U(q)$ is the ground state energy of a q-state Potts model with Gaussian couplings of zero mean and variance J^2. For large q, one can approximate $U(q) = \sqrt{q \ln q}$. The exact formula for calculating $U(q)$ is [17]

$$U(q) = -\frac{1}{4}(q + 1)a + \frac{1}{a}\ln\left[\left(\frac{q}{2\pi}\right)^{1/2}2^{-q+1}\right.$$
$$\left. \times \int_{-\infty}^{\infty} dt \exp\left(-\frac{t^2}{2q} + at\right)\left(1 + \operatorname{erf}\left(\frac{t}{\sqrt{2q}}\right)\right)^{q-1}\right]. \tag{4.14}$$

$U(q)$ must be maximal with respect to the parameter a. This can be evaluated numerically and Table 4.1 gives the results for a few values of q.

One sees that maximum modularity is obtained at $q = 5$, though the value of $U(q)/q$ for $q = 4$ is not much different from it. This qualitative behavior,

Table 4.1. Values of $U(q)/q$ for various values of q obtained from (4.14), which can be used to approximate the expected modularity with (4.13).

q	2	3	4	5	6	7	8	9	10
$U(q)/q$	0.384	0.464	0.484	0.485	0.479	0.471	0.461	0.452	0.442

that dense random graphs tend to cluster into only a few large communities, is confirmed by numerical experiments. By rewriting $M = pN^2/2$ and under the assumption of $p_{ij} = p$ as in the case of Erdős Rényi (ER) random graphs [20], one can further simplify (4.13) and write for the maximum value of the modularity of an ER random graph with connection probability p and N nodes:

$$Q = 0.97\sqrt{\frac{1-p}{pN}}, \tag{4.15}$$

where the fact that $q = 5$ makes the modularity maximal has been used. Figure 4.9 shows the comparison of (4.15) and experiments where the modularity was maximized numerically using a simulated annealing approach as described in an earlier section. One sees that the prediction fits the data well for dense graphs and that modularity decays as a function of $(pN)^{-1/2}$ instead of $(2/pN)^{2/3}$ as proposed in Ref. [19].

Even though the estimations of the value of modularity for random graphs from the Potts spin glass are rather close to the actual situation for sparse random graphs, the number of communities at which maximum modularity is achieved is not. In Ref. [19] it had already been shown that the number of communities for which the modularity reaches a maximum is $\sqrt{N}$ for tree-like networks with $\langle k \rangle = 2$. Unfortunately, no plot was given for the number of communities found in denser networks. The numerical experiments on large Erdős–Rényi random graphs also show that the number of communities found in sparse networks tends to increase as $\langle k \rangle$ decreases.

In general, recursive bi-partitioning will not lead to an optimal community assignment (compare Sect. 4.1.2), shall still be used here for random graphs. It was shown that maximum modularity for random graphs is achieved for

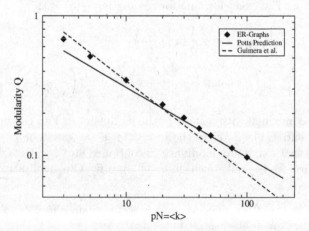

Fig. 4.9. Modularity of Erdős Rényi random graphs with average connectivity $pN = \langle k \rangle$ compared with the estimation from (4.15) and that of Guimera et al. [19]. For the experiment, random graphs with $N = 10,000$ were used.

equipartitions. Hence, one should be able to partition the network recursively and at least find the number of communities in a random graph, for which further partitioning does not result in an improvement of the modularity. The number of cut edges $\mathcal{C} = \mathcal{C}(N, M)$ in any partition will be a function of the number of nodes in the remaining part and the number of connections within this remaining part and their distribution. Note that the M connections will be distributed into internal and external links per node $k_{in} + k_{out} = k$. This allows us to write $\mathcal{C} = N\langle k_{out}\rangle/2$ for a bi-partition. After each partition, the number of internal connections a node has decreases due to the cut. These results are used in order to approximate the number of cut edges after b recursive bi-partitions which lead to 2^b parts:

$$\mathcal{C} = \sum_{t=1}^{b} 2^{t-1} \frac{N}{2^t} \langle k_{out,t}\rangle = \sum_{t=1}^{b} \frac{N}{2} \langle k_{out,t}\rangle, \tag{4.16}$$

where $\langle k_{out,t}\rangle$ is the average number of external edges a node gains after cut t. Since for an Ising model, the ground state energy is $-E_{GS} = M - 2\mathcal{C}$ one finds

$$\frac{\langle k\rangle}{2} + E_{GS}(\langle k\rangle) = \langle k_{out}\rangle. \tag{4.17}$$

Since $k = k_{in} + k_{out}$, one also has

$$\frac{\langle k\rangle}{2} - E_{GS}(\langle k\rangle) = \langle k_{in}\rangle. \tag{4.18}$$

This shows that for any bi-partition, one can always satisfy more than half of the links of every node on average. This also means that any bi-partition will satisfy the definition of community given by Radicchi et al. [8] at least on average, which further means that every random graph has – at least on average – a community structure, assuming Radicchi's definition of community in a strong sense ($k_{in} > k_{out}$) for every node of the random graph. The definition of community in a weak sense $\sum_i k_i^{in} > \sum_i k_i^{out}$ can always be fulfilled in a random graph.

From (4.17) and (4.18) one can then calculate the total number of edges cut after t recursions according to (4.16). One way of doing this is to go back to the results of Fu and Anderson [16] again, who find for a bi-partition

$$\mathcal{C} = \frac{M}{2}\left[1 - c\sqrt{\frac{1-p}{pN}}\right], \tag{4.19}$$

with a constant of $c = 1.5266 \pm 0.0002$ [21]. One can write

$$\langle k_{out}\rangle = \frac{pN - c\sqrt{pN(1-p)}}{2}, \tag{4.20}$$

$$\langle k_{in}\rangle = \frac{pN + c\sqrt{pN(1-p)}}{2}, \tag{4.21}$$

from which one can calculate (4.16) substituting pN with the appropriate $\langle k_{in} \rangle$ in every step of the recursion. The modularity can then be written

$$Q = \frac{2^b - 1}{2^b} - \frac{1}{\langle k \rangle} \sum_{t=1}^{b} \langle k_{out,t} \rangle. \tag{4.22}$$

Now one only needs to find the number of recursions b that maximizes Q. Since the optimal number of recursions will depend on pN, one also finds an estimation of the number of communities in the network.

Figure 4.10 shows a comparison between the theoretical prediction of the maximum modularity that can be obtained from (4.22). The improvement of (4.22) over (4.15) is most likely due to the possibility of having larger numbers of communities, since (4.19) also assumes a Gaussian distribution of local fields, which is a rather poor approximation for the sparse graphs under study. Again, one finds that the modularity behaves asymptotically like $\langle k \rangle^{-1/2}$ as already predicted from the Potts spin glass and contrary to the estimation in [19].

Figure 4.11 shows the comparison of the number of communities estimated from (4.22) and the numerical experiments on random graphs. The good agreement between experiment and prediction is interesting, given the fact that (4.22) allows only powers of two as the number of communities. For dense graphs, the Potts limit of only a few communities is recovered. One observes that sparse random graphs cluster into a large number of communities, while dense random graphs cluster into only a handful of large communities. Most importantly, sparse random graphs exhibit very large values of modularity. These large values are only due to their sparseness and *not* due to small size. It should also be stressed that statistically significant modularity

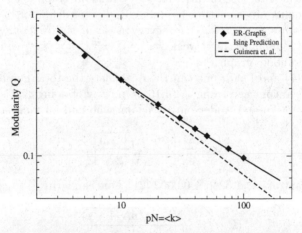

Fig. 4.10. Modularity of Erdős–Rényi random graphs with average connectivity $pN = \langle k \rangle$ compared with the estimation from (4.22) and from Guimera et al. [19]. For the experiment, random graphs with $N = 10,000$ were used.

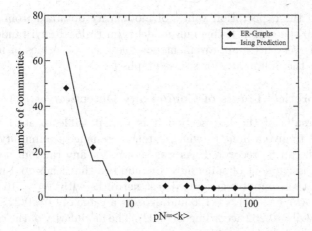

Fig. 4.11. Number of communities found in Erdős–Rényi random graphs with average connectivity $pN = \langle k \rangle$ compared with the estimation from (4.22). For the experiment, random graphs with $N = 10,000$ nodes were used.

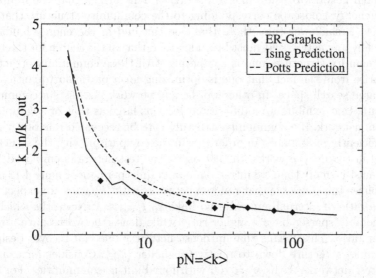

Fig. 4.12. Ratio of internal links to external links k_{in}/k_{out} in the ground state of the Hamiltonian. Shown are the experimental values from clustering random graphs with $N = 10,000$ nodes and the expectation values calculated from using a Potts model (4.15) or an Ising model (4.22) recursively. The dotted line represents the Radicchi et al. definition of community in a "strong sense" [8]. Note that sparse graphs will, on average, always exhibit such communities, while dense graphs will not, even though their modularity may be well above the expectation value for an equivalent random graph.

must exceed the expectation values of modularity obtained from a suitable null model of the graph. If this null model is an Erdős–Rényi random graph, then there is very little improvement possible over the values of modularity obtained for the null model for sparse graphs.

4.4.2 Theoretical Limits of Community Detection

With the results of the last section it is now possible to start explaining Fig. 4.5 and to give a limit to which extent a designed community structure in a network can be recovered. As was shown, for any random network one can find an assignment of spins into communities that leads to a modularity $Q > 0$. For the computer-generated test networks with $\langle k \rangle = 16$ one has a value of $p = \langle k \rangle/(N-1) = 0.126$ and expects a value of $Q = 0.227$ according to (4.15) and $Q = 0.262$ according to (4.22). The modularity of the community structure built in by design is given by

$$Q(\langle k_{in} \rangle) = \frac{\langle k_{in} \rangle}{\langle k \rangle} - \frac{1}{4} \tag{4.23}$$

for a network of four equal sized groups of 32 nodes. Hence, below $\langle k_{in} \rangle = 8$, one has a designed modularity that is smaller than what can be expected from a random network of the same connectivity! This means that the minimum in the energy landscape corresponding to the community structure that was designed is shallower than those that one can find in the energy landscape defined by any network. It must be understood that in the search for the built-in community structure, one is competing with those community structures that arise from the fact that one is optimizing for a particular quantity in a very large search space. In other words, any network possesses a community structure that exhibits a modularity at least as large as that of a completely random network. If a community structure is to be recovered reliably, it must be sufficiently pronounced in order to win the comparison with the structures arising in random networks. In the case of the test networks employed here, there must be more than ≈ 8 intra-community links per node. Figure 4.12 again exemplifies this. Observe that random networks with $\langle k \rangle = 16$ are expected to show a ratio of internal and external links $k_{in}/k_{out} \approx 1$. Networks which are considerably sparser have a higher ratio while denser networks have a much smaller ratio. This means that in dense networks one can recover designed community structure down to relatively smaller $\langle k_{in} \rangle$. Consider for example large test networks with $\langle k \rangle = 100$ with four built-in communities. For such networks one expects a modularity of $Q \approx 0.1$ and hence the critical value of intra-community links to which the community structure could reliably be estimated would be $\langle k_{in} \rangle_c = 35$ which is much smaller in relative comparison to the average degree in the network.

This also means that the point at which one cannot distinguish between a random and a modular network is not defined by $p_{in} = p_{out} = p$ for the internal and external link densities as one may have intuitively expected.

Rather, it is determined by the ratio of $\langle k_{in}\rangle/(\langle k\rangle - \langle k_{in}\rangle)$ in the ground state of a random network and depends on the connectivity of the network $\langle k\rangle$.

Finally, from Fig. 4.12 one observes that sparse random graphs all show communities in the strong sense of Radicchi et al. [8] . Further, it is very difficult to find communities in the strong sense in dense graphs, even though they may exhibit a modularity well above that of a random graph.

4.5 Conclusion

In this chapter we focussed on modular or community structures in networks. We have seen how they naturally form a sub-class of the block models introduced in the previous chapter. By recognizing the formal equivalence of the quality function for a diagonal block structure with the energy of an infinite range spin glass, we could derive first estimates for the expectation value of the fit of an entirely random network to a diagonal block model. It was shown that any structure present in a network competes with spurious structures that arise due to random fluctuations in the link structure of the network and the optimization process carried out by the community detection algorithm. Such competition may render a community structure in a network undetectable. Note that this is not a finite size effect which vanishes in the thermodynamic limit but rather persists at all scales. The following two chapters will address this problem in greater detail.

References

1. M. E. J. Newman and M. Girvan. Finding and evaluating community structure in networks. *Physical Review E*, 69:026113, 2004.
2. A. Clauset, M. E. J. Newman, and C. Moore. Finding community structure in very large networks. *Physical Review E*, 70:066111, 2004.
3. M. R. Gareya and D. S. Johnson. Computers and Intractability. W. H. Freeman & Co Ltd, London, 1979.
4. A. K. Jain, M. N. Murty, and P. J. Flynn. Data clustering: A review. *ACM Computing Surveys*, 31(3):264–323, 1999.
5. M. E. J. Newman. Fast algorithm for detecting community structure in networks. *Physical Review E*, 69:066133, 2004.
6. S. Wasserman and K. Faust. *Social Network Analysis*. Cambridge University Press, New York, 1994.
7. S. B. Seidmann. Internal cohesion of ls-sets in graphs. *Social Networks*, 5:97–107, 1983.
8. F. Radicchi, C. Castellano, F. Cecconi, V. Loreto, and D. Parisi. Defining and identifying communities in networks. *Proceedings of the National Academy of Sciences of the United States of America*, 101:2658, 2004.
9. I. Derényi, G. Palla, and T. Vicsek. Clique percolation in random networks. *Physical Review Letters*, 94:160202, 2005.

10. G. Palla, I. Derenyi, I. Farkas, and T. Vicsek. Uncovering the overlapping community structure of complex networks in nature and society. *Nature*, 435:814, 2005.

11. A. Arenas, A. D'iaz-Guilera, and C. J. Pérez-Vicente. Synchronization reveals topological scales in complex networks. *Physical Review Letters*, 96:114102, 2006.

12. D. Gfeller, J. -C. Chappelier, and P. de los Rios. Finding instabilities in the community structure of complex networks. *Physical Review E*, 72:056135, 2005.

13. R. Palmer. Statistical mechanics approaches to complex optimization problems. In P. W. Anderson, K. J. Arrow, and D. Pines, editors, *The Economy as an Evolving Complex System*, chapter II, p. 177. Westview-Press, New York, 1988.

14. M. Girvan and M. E. J. Newman. Community structure in social and biological networks. *Proceedings of the National Academy of Sciences of the United States of America*, 99(12):7821–7826, 2002.

15. S. Maslov and K. Sneppen. Specificity and stability in topology of protein networks. *Science*, 296:910–913, 2002.

16. Y. Fu and P. W. Anderson. Application of statistical mechanics to NP-complete problems in combinatorial optimisation. *Journal of Physics A: Mathematical and General*, 19:1605–1620, 1986.

17. I. Kanter and H. Sompolinsky. Graph optimisation problems and the Potts glass. *Journal of Physics A: Mathematical and General*, 20:L636–679, 1987.

18. M. Mezard, G. Parisi, and M. A. Virasoro. *Spin Glass Theory and Beyond*. World Scientific, Singapore, 1987.

19. R. Guimera, M. Sales-Pardo, and L. N. Amaral. Modularity from fluctuations in random graphs and complex networks. *Physical Review E*, 70:025101(R), 2004.

20. P. Erdős and A. Rényi. On the evolution of random graphs. *Publications of the Mathematical Institute of the Hungarian Academy of Sciences*, 5:17–61, 1960.

21. J. R. Banavar, D. Sherrington, and N. Sourlas. Graph bipartioning and statistical mechanics. *Journal of Physics A: Mathematical and General*, 20:L1–L8, 1987.

5

Modularity of Dense Random Graphs

In the last chapter, it was shown how the problem of community detection
can be mapped onto finding the ground state of an infinite range spin glass.
Further, it was demonstrated that for random graphs, maximum modularity
is achieved for an equipartition due to entropic reasons. It was possible to use
known results from graph partitioning to give estimates of the modularity in
ER random graphs. However, these results only apply to dense graphs with a
Poissonian degree distribution. In this chapter, the ground state energy of the
modularity Hamiltonian will be calculated directly for any degree distribution.
The entire development will follow closely along the lines of Fu and Anderson
(FA) [1, 2].

5.1 Analytical Developments

Let us recall the modularity Hamiltonian:

$$\mathcal{H} = -\sum_{i<j}(A_{ij} - \gamma p_{ij})\delta(\sigma_i, \sigma_j). \tag{5.1}$$

For convenience, instead of a Potts model with q different spin states, the
discussion is limited to only two spin states as in the Ising model, namely
$S_i \in -1, 1$. The delta function in (5.1) can be expressed as

$$\delta(S_i, S_j) = \frac{1}{2}S_iS_j + \frac{1}{2}, \tag{5.2}$$

which leads to the new Hamiltonian

$$\mathcal{H} = -\sum_{i<j}(A_{ij} - \gamma p_{ij})S_iS_j. \tag{5.3}$$

Note that (5.3) differs from (5.1) only by an irrelevant constant which even
vanishes for $\gamma = 1$ due to the normalization of p_{ij}. Because of the factor $1/2$

Reichardt, J.: *Modularity of Dense Random Graphs*. Lect. Notes Phys. **766**, 69–86 (2009)
DOI 10.1007/978-3-540-87833-9_5   © Springer-Verlag Berlin Heidelberg 2009

in (5.2), the modularity of the partition into two communities is now and for the remainder of this chapter

$$Q_2 = -\frac{\mathcal{H}}{2M},\tag{5.4}$$

where $\mathcal{H}$ now denotes the Hamiltonian (5.3). For the number of cut edges of the partition one can write

$$\mathcal{C} = \frac{1}{2}(M + E_g) = \frac{M}{2}(1 - 2Q_2),\tag{5.5}$$

with E_g denoting the ground state energy of (5.3) and it is clear that Q_2 measures the improvement of the partition over a random assignment into groups.

Formally, (5.3) corresponds to a Sherrington–Kirkpatrick (SK) model of a spin glass [3]

$$\mathcal{H} = -\sum_{i<j} J_{ij} S_i S_j,\tag{5.6}$$

with couplings of the form

$$J_{ij} = (A_{ij} - \gamma p_{ij}).\tag{5.7}$$

Different from the SK model, however, the couplings are not drawn from a symmetric distribution, but there exist a few strong ferromagnetic couplings along the links of the network and many anti-ferromagnetic couplings between unconnected nodes. It is convenient to differentiate between the two and define

$$J_{ij}^+ = 1 - \gamma p_{ij}, \qquad J_{ij}^- = -\gamma p_{ij} \qquad \text{and} \qquad J = J_{ij}^+ - J_{ij}^- = 1.\tag{5.8}$$

Note that $J = 1$ regardless of the choice of connection model and the parameter γ. The coupling distribution q_{ij} is now determined completely by the ensemble of networks we are considering. Recall that connection model p_{ij} describes the probability for an edge to be absent or present in a particular realization of the network and hence parameterizes an ensemble of networks.

The probability density function of the coupling distribution $q_{ij}(J_{ij})$ between two nodes i and j can then be written as

$$q_{ij}(J_{ij}) = p_{ij}\delta(J_{ij} - J_{ij}^+) + (1 - p_{ij})\delta(J_{ij} - J_{ij}^-).\tag{5.9}$$

Note the mean of this distribution is determined by γ and the average value of p_{ij}:

$$[J_{ij}] = p_{ij}J_{ij}^+ + (1 - p_{ij})J_{ij}^- = (1 - \gamma)p_{ij},\tag{5.10}$$

where the symbol $[\cdot]$ denotes an average over the graph ensemble. For the variance of q_{ij}, we find

$$\sigma_{ij}^2 = p_{ij}(J_{ij}^+ - [J_{ij}])^2 + (1 - p_{ij})(J_{ij}^- - [J_{ij}])^2 = p_{ij}(1 - p_{ij}),\tag{5.11}$$

which is independent of γ.

In order to deal with uncorrelated graphs of arbitrary degree distribution, a connection model of the form

$$p_{ij} = \frac{g_i g_j}{\langle g \rangle N} \tag{5.12}$$

is chosen, i.e., we assume that the probability of two nodes being connected depends only on the product of some variable g_i associated with each node i, a trick that has been introduced in Ref. [4]. The distribution $p(g)$ of the g_i is supposed to be known. We further assume that the g_i are such that $[J_{ij}]$ and $[\sigma_{ij}^2]$ (to leading order) scale as $1/N$.

Note that due to the normalization condition on the connection model $\sum_{ij} p_{ij} = \langle k \rangle N$ we have two conditions on $p(g)$. The first is that $\sum_i g_i = \sqrt{\langle k \rangle \langle g \rangle} N$ from which directly follows that $\langle g \rangle = \sqrt{\langle k \rangle \langle g \rangle}$. Second, the variance of q_{ij} must be positive for all pairs of nodes which requires that $g_i g_j < \langle g \rangle N$. This condition translates into the fact that $\langle g^2 \rangle < \langle g \rangle N$, i.e., the second and higher moments of g do not diverge faster than the network size. Both of these observations and the fact that p_{ij} factorizes are crucial for the following developments. An obvious choice is of course to set $g_i = k_i$, i.e., to assume the vertex weights equal to the degrees of the vertices. Then, the first condition on $p(g)$ is trivially fulfilled. Provided that the degree distribution also fulfills the second requirement, the following results are valid.

The goal is now to calculate the ground state energy E_g averaged over the ensemble of graphs with a given degree distribution, i.e., those described by the connection model p_{ij} and a given degree distribution. From $F = -\beta^{-1} \ln Z$ one sees that it is necessary to calculate the logarithm of the partition function Z. Instead of doing so directly it is easier to use the replica trick [3,5]:

$$[\ln Z] = \lim_{n \to 0} \frac{[Z^n] - 1}{n}, \tag{5.13}$$

where the symbol $[\cdot]$ denotes an average over the graph ensemble. So instead of calculating the logarithm of the partition function, it is only necessary to calculate the n-th power of it which turns out to be analytically tractable. The n-th power is written as

$$Z^n = \mathrm{Tr}_n \exp\left\{ \beta \sum_{\alpha}^{n} \sum_{i<j} J_{ij} S_i^\alpha S_j^\alpha \right\}. \tag{5.14}$$

Here Tr_n denotes the trace over the n replicated spins, i.e.,

$$\mathrm{Tr}_n = \sum_{S_1^1 = -1}^{1} \sum_{S_1^2 = -1}^{1} \cdots \sum_{S_1^n = -1}^{1} \cdots \sum_{S_N^1 = -1}^{1} \sum_{S_N^2 = -1}^{1} \cdots \sum_{S_N^n = -1}^{1}. \tag{5.15}$$

The average over the graph ensemble is now, with the help of (5.9), written as

$$[Z^n] = \int \prod_{i<j} (dJ_{ij} q_{ij}(J_{ij})) \, \mathrm{Tr}_n \exp\left\{ \beta \sum_{\alpha}^{n} \sum_{i<j} J_{ij} S_i^\alpha S_j^\alpha \right\}. \tag{5.16}$$

Each of the integrals can be drawn into the trace and by rewriting the sum in the exponential also as a product, this can be reformulated as

$$[Z^n] = \text{Tr}_n \prod_{i<j} \int \exp\left\{\beta \sum_\alpha^n J_{ij} S_i^\alpha S_j^\alpha\right\} q_{ij}(J_{ij}) dJ_{ij}. \qquad (5.17)$$

Using (5.9) to perform the integrals this is reduced to

$$[Z^n] = \text{Tr}_n \prod_{i<j} \left[p_{ij} \exp\left\{\beta J_{ij}^+ \sum_\alpha^n S_i^\alpha S_j^\alpha\right\} + (1-p_{ij}) \exp\left\{\beta J_{ij}^- \sum_\alpha^n S_i^\alpha S_j^\alpha\right\}\right], \qquad (5.18)$$

which is conveniently rewritten as

$$[Z^n] = \text{Tr}_n \prod_{i<j} (1-p_{ij}) \exp\left\{\beta J_{ij}^- \sum_\alpha^n S_i^\alpha S_j^\alpha\right\}$$

$$\times \left[1 + \frac{p_{ij}}{1-p_{ij}} \exp\left\{\beta(J_{ij}^+ - J_{ij}^-) \sum_\alpha^n S_i^\alpha S_j^\alpha\right\}\right]. \qquad (5.19)$$

Now $p_0 = \frac{p_{ij}}{1-p_{ij}}$ is defined, keeping in mind that it depends on both i and j. Further, recalling $J = J_{ij}^+ - J_{ij}^- = 1$, expression (5.19) is rewritten using the identity $\prod a = \exp \ln \prod a = \exp \sum \ln a$ as

$$[Z^n] = \text{Tr}_n \exp\left[\sum_{i<j} \ln(1-p_{ij}) + \beta J_{ij}^- \sum_\alpha^n S_i^\alpha S_j^\alpha \right.$$

$$\left. + \ln\left(1 + p_0 \exp\left\{\beta J \sum_\alpha^n S_i^\alpha S_j^\alpha\right\}\right)\right]. \qquad (5.20)$$

Let us concentrate on the last term involving $\ln(1 + p_0 \exp)$ and write the logarithm and exponential as a series. One finds

$$\ln\left(1 + p_0 \exp\left\{\beta J \sum_\alpha^n S_i^\alpha S_j^\alpha\right\}\right) = \sum_{l=1}^\infty \frac{-1^{l-1}}{l} \left(p_0 \exp\left\{\beta J \sum_\alpha^n S_i^\alpha S_j^\alpha\right\}\right)^l \qquad (5.21)$$

for the logarithm and

$$\exp\left\{\beta J \sum_\alpha^n S_i^\alpha S_j^\alpha\right\}^l = \sum_{k_1=0}^\infty \sum_{k_2=0}^\infty \cdots \sum_{k_l=0}^\infty \frac{(\beta J)^{k_1+k_2+\cdots+k_l}}{k_1! k_2! \ldots k_l!} \left(\sum_\alpha^n S_i^\alpha S_j^\alpha\right)^{k_1+k_2+\cdots+k_l} \qquad (5.22)$$

for the exponential. Putting expression (5.22) and (5.21) together and regrouping the terms according to the exponent of βJ, one finds for $k_1+k_2+\cdots+k_l = 0$ the following coefficient:

$$\sum_{l=1}^{\infty} \frac{-1^{l-1}}{l} p_0^l = \ln(1 + p_0). \tag{5.23}$$

For $(\beta J)^1$, i.e., for $k_1 + k_2 + \cdots + k_l = 1$, there exist l possibilities for each of the k_m to be non-zero. Hence, for the second coefficient one finds

$$\beta J \sum_{\alpha}^{n} S_i^\alpha S_j^\alpha \sum_{l=1}^{\infty} \frac{-1^{l-1}}{l} l p_0^l. \tag{5.24}$$

For $(\beta J)^2$, i.e., $k_1 + k_2 + \cdots + k_l = 2$, there exist $l(l-1)/2$ possibilities for 2 of the k_m to be one and the rest zero and there exist l possibilities for one of the k_n to be two. In the last case, the $k_m!$ terms introduce a factor of $1/2$ and one is left with $l^2/2$ terms with βJ raised to the power of 2. Therefore, the third coefficient is

$$(\beta J)^2 \left(\sum_{\alpha}^{n} S_i^\alpha S_j^\alpha \right)^2 \frac{1}{2} \sum_{l=1}^{\infty} \frac{-1^{l-1}}{l} l^2 p_0^l. \tag{5.25}$$

Note that with the help of the geometric series $\sum_{n=1}^{\infty} q^n = \frac{q}{1-q}$ in (5.24), one can write

$$\sum_{l=1}^{\infty} \frac{-1^{l-1}}{l} l p_0^l = \sum_{l=1}^{\infty} (p_0^{2l-1}) - \sum_{l=1}^{\infty} p_0^{2l} = \left(\frac{1}{p_0} - 1 \right) \sum_{l=1}^{\infty} \left(p_0^2 \right)^l$$

$$= \left(\frac{1 - p_0}{p_0} \right) \left(\frac{p_0^2}{1 - p_0^2} \right) = \frac{p_0}{1 + p_0} = p_{ij}. \tag{5.26}$$

For the second coefficient from (5.25) one can write with the help of (5.24)

$$\sum_{l=1}^{\infty} \frac{-1^{l-1}}{l} l^2 p_0^l = p_0 \frac{\partial}{\partial p_0} \sum_{l=1}^{\infty} \frac{-1^{l-1}}{l} l p_0^l = p_0 \frac{\partial}{\partial p_0} \frac{p_0}{1 + p_0}$$

$$= \frac{p_0}{(1 + p_0)^2} = p_{ij}(1 - p_{ij}). \tag{5.27}$$

Using (5.23), (5.26) and (5.27) one can rewrite the term involving $\ln(1 + p_0 \exp)$ from (5.20) in the following way:

$$\ln(1 + p_0) + \beta J p_{ij} \sum_{\alpha}^{n} S_i^\alpha S_j^\alpha + \frac{(\beta J)^2}{2} p_{ij}(1 - p_{ij}) \left(\sum_{\alpha}^{n} S_i^\alpha S_j^\alpha \right)^2. \tag{5.28}$$

Higher orders of (βJ) have all been dropped. The full expression for the partition function (5.20) is then written as

$$[Z^n] = \text{Tr}_n \exp \left[\sum_{i<j} \ln(1 - p_{ij}) + \ln(1 + p_0) \right.$$

$$\left. + \beta \left(J_{ij}^- + p_{ij} J \right) \sum_\alpha S_i^\alpha S_j^\alpha + \frac{(\beta J)^2}{2} p_{ij}(1 - p_{ij}) \left(\sum_\alpha^n S_i^\alpha S_j^\alpha \right)^2 \right].$$

$$(5.29)$$

Keeping in mind the definition of $p_0 = \frac{p_{ij}}{1-p_{ij}}$ the two ln terms cancel and (5.29) reduces to

$$[Z^n] = \text{Tr}_n \exp \left[\sum_{i<j} \beta \left(J_{ij}^- + p_{ij} J \right) \sum_\alpha S_i^\alpha S_j^\alpha + \frac{(\beta J)^2}{2} p_{ij}(1 - p_{ij}) \left(\sum_\alpha^n S_i^\alpha S_j^\alpha \right)^2 \right].$$

$$(5.30)$$

Until now, the presented treatment has followed Ref. [1] almost one to one. Note how the mean and the variance of the coupling distribution have entered the calculation. We could have arrived at this point also by assuming a Gaussian coupling distribution of the same mean and variance.

At this point, the development deviates, however. The first and second addend in (5.30) will be treated separately and the factorization of $p_{ij} = g_i g_j / \langle g \rangle N$ is used for the first time:

$$\sum_{i<j} \beta \left(J_{ij}^- + p_{ij} J \right) \sum_\alpha S_i^\alpha S_j^\alpha = \beta(J - \gamma) \sum_\alpha \sum_{i<j} p_{ij} S_i^\alpha S_j^\alpha$$

$$= \frac{\beta(J - \gamma)}{N\langle g \rangle} \sum_\alpha \frac{1}{2} \left[\left(\sum_i g_i S_i^\alpha \right)^2 - \sum_i g_i^2 \right]$$

$$= \frac{\beta(J - \gamma)}{2N\langle g \rangle} \sum_\alpha \left(\sum_i g_i S_i^\alpha \right)^2 - \frac{\beta(J - \gamma)n\langle g^2 \rangle}{2}.$$

$$(5.31)$$

The last term vanishes in the limit $n \to 0$. For the second addend in (5.30) one finds by neglecting p_{ij}^2 vs. p_{ij}

$$\sum_{i<j} \frac{(\beta J)^2}{2} (p_{ij} - p_{ij}^2) \left(\sum_\alpha^n S_i^\alpha S_j^\alpha \right)^2 = \frac{(\beta J)^2}{2} \sum_{i<j} p_{ij} \left[\sum_{\alpha \neq \beta} S_i^\alpha S_j^\beta S_j^\alpha S_j^\beta + n \right]$$

$$= \frac{(\beta J)^2}{2N} \left[n \sum_{i<j} g_i g_j \right.$$

$$\left. + \sum_{\alpha \neq \beta} \frac{1}{2} \left(\left(\sum_i g_i S_i^\alpha S_i^\beta \right)^2 - \sum_i g_i^2 \right) \right]$$

$$= \frac{(\beta J)^2}{2N\langle g \rangle} \sum_{\alpha < \beta} \left(\sum_i g_i S_i^\alpha S_i^\beta \right)^2 + n \frac{(\beta J)^2 N\langle g \rangle}{4}.$$

$$(5.32)$$

In the last line, the term proportional to n^2 was dropped as it is much smaller than the term proportional to n in the limit $n \to 0$. With this, the partition function can be written in a form which can be reduced by Gaussian integrals:

$$[Z^n] = \exp\left(\frac{(\beta J)^2 n \langle g \rangle N}{4}\right) \mathrm{Tr}_n \exp\left\{\frac{\beta(J-\gamma)}{2N\langle g \rangle} \sum_\alpha \underbrace{\left(\sum_i g_i S_i^\alpha\right)^2}_{m_\alpha^*}\right.$$

$$\left. + \frac{(\beta J)^2}{2N\langle g \rangle} \sum_{\alpha < \beta} \underbrace{\left(\sum_i g_i S_i^\alpha S_i^\beta\right)^2}_{q_{\alpha\beta}^*}\right\}. \qquad (5.33)$$

This is formally equivalent to the SK model [3] except for the g_i in the sums over spins. By using a Hubbard–Stratonovich identity [6]

$$\exp x^2 = \int_{-\infty}^\infty dm \exp(-\pi m^2 - 2\sqrt{\pi}xm) \qquad (5.34)$$

in the following form

$$\exp\left(\frac{a}{2}x^2\right) = \sqrt{\frac{aN\langle g \rangle}{2\pi}} \int_{-\infty}^\infty dm \exp\left(-\frac{aN\langle g \rangle}{2}m^2 + \sqrt{N\langle g \rangle}axm\right), \qquad (5.35)$$

the terms of (5.33) are simplified to give

$$\exp\left\{\frac{a}{2N\langle g \rangle} \sum_\alpha \left(\sum_i g_i S_i^\alpha\right)^2\right\} = \prod_\alpha \sqrt{\frac{aN\langle g \rangle}{2\pi}} \int dm_\alpha \exp\left\{-\frac{aN\langle g \rangle}{2}m_\alpha^2\right.$$

$$\left. + am_\alpha \sum_i g_i S_i^\alpha\right\}, \qquad (5.36)$$

where the abbreviation $a = \beta(J-\gamma)$ is used and x is set to $x = \sum_i g_i S_i^\alpha / \sqrt{N\langle g \rangle}$. Further, one finds

$$\exp\left\{\frac{b}{2} \sum_{\alpha < \beta} \left(\sum_i g_i S_i^\alpha S_i^\beta\right)^2\right\} = \prod_{\alpha < \beta} \sqrt{\frac{bN\langle g \rangle}{2\pi}} \int dq_{\alpha\beta} \exp\left\{-\frac{bN\langle g \rangle}{2}q_{\alpha\beta}^2\right.$$

$$\left. + bq_{\alpha\beta} \sum_i g_i S_i^\alpha S_i^\beta\right\}, \qquad (5.37)$$

with $b = (\beta J)^2$. The integration variables m_α and $q_{\alpha\beta}$ introduced here are called the order parameters of the system and turn out to have a profound physical meaning. Let us now focus on what is left under the trace after the transformation using Gaussian integrals:

$$\text{Tr}_n \exp\left\{\sum_i \left[ag_i \sum_\alpha S_i^\alpha m_\alpha + bg_i \sum_{\alpha<\beta} S_i^\alpha S_i^\beta q_{\alpha\beta}\right]\right\}. \tag{5.38}$$

Recalling the fact that the g_i describe the nodes, the sum over the nodes can be replaced by a sum over the distribution of the g_i

$$\sum_i^N g_i = N \sum_k p(g_k)g_k = \sum_k n_k g_k, \tag{5.39}$$

where n_k is the number of nodes that have vertex weight g_k and $p(g_k)$ is the probability that a randomly drawn vertex has vertex weight g_k. Hence, we can write

$$\prod_k \left(\text{Tr}_n \exp\left\{\underbrace{\beta(J-\gamma)g_k \sum_\alpha S^\alpha m_\alpha + (\beta J)^2 g_k \sum_{\alpha<\beta} S^\alpha \dot{S}^\beta q_{\alpha\beta}}_{L_k}\right\}\right)^{n_k}. \tag{5.40}$$

With this, the trace over an N-site problem from (5.38) has been reduced to the product of single site traces, one for each type of node with weight g_k. The symbol Tr_n now represents only the trace over n replicas at a single site. Again (5.40) is written as an exponential in order to use it for the calculation of the partition function. Leaving off all the factors which vanish when taking the limit n to zero, one arrives at

$$[Z^n] = \exp\left(\frac{(\beta J)^2 n\langle g\rangle N}{4}\right) \int \prod_\alpha dm_\alpha \int \prod_{\alpha<\beta} dq_{\alpha\beta}$$

$$\times \exp\left\{-\frac{\beta(J-\gamma)\langle g\rangle N}{2}\sum_\alpha m_\alpha^2 - \frac{(\beta J)^2\langle g\rangle N}{2}\sum_{\alpha<\beta} q_{\alpha\beta}^2 + N\sum_k p(g_k)\ln\text{Tr}_n\exp(L_k)\right\}. \tag{5.41}$$

The integrand is now of the following form:

$$\int dm \exp(-Nh(m)). \tag{5.42}$$

Expanding $h(m)$ around its maximum value at m^* one has $h(m^* + \Delta m) = h(m^*) + \mathcal{O}(\Delta m^2)$ and thus

$$\int dm \exp(-Nh(m)) \approx \exp(-Nh(m^*)). \tag{5.43}$$

This method of approximating the integral by the maximum value of the integrand is called "steepest descent" or "saddle point" method [7]. Applying this method to (5.41) leads to

$$[Z^n] \approx \exp \left\{ -\frac{\beta(J-\gamma)\langle g\rangle N}{2} \sum_\alpha m_\alpha^{*2} - \frac{(\beta J)^2 \langle g\rangle N}{2} \sum_{\alpha<\beta} q_{\alpha\beta}^{*2} \right.$$

$$\left. +N \sum_k p(g_k) \ln \mathrm{Tr_n} \exp(L_k) + \frac{(\beta J)^2 n\langle g\rangle N}{4} \right\}. \quad (5.44)$$

Here, m_α^* and $q_{\alpha\beta}^*$ denote the respective values which maximize the integrand. Next (5.44) is expanded around zero. This is possible by assuming that the limit $n \to 0$ is taken at fixed N and only after this, the limit $N \to \infty$ is taken. Keeping only the first order, one arrives at

$$[Z^n] \approx 1 + nN \left\{ -\frac{\beta(J-\gamma)\langle g\rangle}{2n} \sum_\alpha m_\alpha^{*2} - \frac{(\beta J)^2 \langle g\rangle}{2n} \sum_{\alpha<\beta} q_{\alpha,\beta}^{*2} \right.$$

$$\left. +\frac{1}{n} \sum_k p(g_k) \ln \mathrm{Tr_n} \exp(L_k) + \frac{(\beta J)^2 \langle g\rangle}{4} \right\}. \quad (5.45)$$

The free energy per spin now becomes $-\beta[f] = [\ln Z] = \lim_{n\to 0}([Z^n]-1)/nN$

$$-\beta[f] = \lim_{n\to 0} \left\{ -\frac{\beta(J-\gamma)\langle g\rangle}{2n} \sum_\alpha m_\alpha^{*2} - \frac{(\beta J)^2 \langle g\rangle}{2n} \sum_{\alpha<\beta} q_{\alpha\beta}^{*2} \right.$$

$$\left. +\frac{1}{n} \sum_k p(g_k) \ln \mathrm{Tr_n} \exp(L_k) + \frac{(\beta J)^2 \langle g\rangle}{4} \right\}. \quad (5.46)$$

Thus far, one cannot determine the values of m_α and $q_{\alpha\beta}$ which make the integrand in (5.41) maximal. However, the free energy must be extremal with respect to all order parameters since these now characterize the system in place of the individual spins:

$$\frac{\partial[f]}{\partial q_{\alpha\beta}} = 0, \quad (5.47)$$

$$\frac{\partial[f]}{\partial m_\alpha} = 0. \quad (5.48)$$

Taking these derivatives in (5.46) explicitly one finds for (5.47)

$$(\beta J)^2 \langle g\rangle q_{\alpha\beta}^* = \sum_k p(g_k) \frac{1}{\mathrm{Tr_n} \exp(L_k)} (\beta J)^2 g_k \mathrm{Tr_n} S^\alpha S^\beta \exp(L_k), \quad (5.49)$$

which is more conveniently written as

$$q_{\alpha\beta}^* = \frac{1}{\langle g\rangle} \langle g_k \langle S^\alpha S^\beta \rangle_{L_k} \rangle_k. \quad (5.50)$$

Here $\langle \cdot \rangle_{L_k}$ is an average with respect to the trace L_k defined in (5.40) and $\langle \cdot \rangle_k$ is an average over the distribution of vertex weights. It now becomes clear that the order parameter $q_{\alpha\beta}^*$ measures the overlap of spin states between different replicas weighted by the degree of the nodes. For m_α^* one finds in the same fashion for (5.48):

$$m_\alpha^* = \frac{1}{\langle g \rangle} \langle g_k \langle S^\alpha \rangle_{L_k} \rangle_k, \tag{5.51}$$

and it is clear that m_α corresponds to the magnetization of the system again weighted by the degree of the nodes.

For the further development one needs to assume some form of dependence of the order parameters of the replica index. The intuitive assumption is independence, i.e., $q_{\alpha\beta}^* = q$ and $m_\alpha^* = m$ independent of the replica index. This assumption is known as the replica symmetric assumption and turns out to be wrong because it leads to an unphysical behavior of the ground state entropy. For reasons of simplicity and since the replica symmetric assumption is still a good approximation for the ground state energy, it is the method of choice at this point. The free energy density is then written as

$$-\beta[f] = \lim_{n \to 0} \left\{ -\frac{\beta(J - \gamma)\langle g \rangle}{2} m^2 - \frac{(\beta J)^2 \langle g \rangle}{4}(n - 1)q^2 \right.$$

$$\left. + \frac{1}{n} \sum_k p(g_k) \ln \mathrm{Tr}_n \exp(L_k) + \frac{(\beta J)^2 \langle g \rangle}{4} \right\}. \tag{5.52}$$

The logarithm of the trace is treated separately:

$$\ln \mathrm{Tr}_n e^{L_k} = \ln \mathrm{Tr}_n \exp \left\{ \beta(J - \gamma)g_k m \sum_\alpha S^\alpha + \frac{(\beta J)^2 g_k}{2} q \left[\left(\sum_\alpha S^\alpha \right)^2 - n \right] \right\}. \tag{5.53}$$

Again it is decomposed by a Hubbard–Stratonovich identity and the following two abbreviations are introduced:

$$q_k = q g_k \quad \text{and} \quad m_k = m g_k. \tag{5.54}$$

One finds

$$\ln \mathrm{Tr}_n e^{L_k} = \ln \mathrm{Tr}_n \sqrt{\frac{(\beta J)^2 q_k}{2\pi}} \int dz \exp \left\{ -\frac{(\beta J)^2 q_k}{2} z^2 + (\beta J)^2 q_k z \sum_\alpha S^\alpha \right.$$

$$\left. -\frac{n(\beta J)^2 q_k}{2} + \beta(J - \gamma)m_k \sum_\alpha S^\alpha \right\}$$

$$= \ln \int Dz \exp \left\{ n \ln \left[2 \cosh\left(\beta J \sqrt{q_k} z + \beta(J - \gamma)m_k\right) - \frac{n(\beta J)^2 q_k}{2} \right] \right\}$$

$$= \ln \left(1 + n \int Dz \ln \left[2 \cosh(\beta \tilde{H}_k(z)) \right] - \frac{n(\beta J)^2 q_k}{2} + \mathcal{O}(n^2) \right). \tag{5.55}$$

Here the Gaussian measure of $Dz = \exp(-z^2/2)/\sqrt{(2\pi)}$ has been introduced which results from completing the square in the integrand. Further one has

$\tilde{H}_k(z) = J\sqrt{q_k}z + (J-\gamma)m_k$. Finally one can write for the free energy density by taking the limit $n \to 0$ and expanding the logarithm of (5.55) around $n = 0$:

$$-\beta[f] = \frac{(\beta J)^2 \langle g \rangle}{4}(1-q)^2 - \frac{\beta(J-\gamma)\langle g \rangle}{2}m^2 + \sum_k p(g_k) \int Dz \ln 2\cosh(\beta\tilde{H}_k(z)).$$

(5.56)

The equations of state for the order parameter m denoting the magnetization is then written as

$$m = \frac{1}{\langle g \rangle} \sum_k p(g_k)g_k \int Dz \tanh(\beta\tilde{H}_k(z)). \qquad (5.57)$$

For the spin glass order parameter q one finds

$$0 = \frac{(\beta J)^2 \langle g \rangle}{2}(q-1) + \sum_k p(g_k)\frac{\beta J\sqrt{g_k}}{2\sqrt{q}} \int Dz \tanh(\tilde{H}_k(z))z. \qquad (5.58)$$

Partial integration then yields

$$\langle g \rangle q = \langle g \rangle - \sum_k p(g_k)g_k \int Dz\,\text{sech}^2(\beta\tilde{H}_k(z))$$

$$= \sum_k p(g_k)g_k \left(1 - \int Dz\,\text{sech}^2(\beta\tilde{H}_k(z))\right)$$

$$q = \frac{1}{\langle g \rangle} \sum_k p(g_k)g_k \int Dz \tanh^2(\beta\tilde{H}_k(z)). \qquad (5.59)$$

These equations of state (5.57) and (5.59) are in exact correspondence with those derived earlier in (5.51) and (5.50).

Let us now study the case of $\gamma = 1$ which corresponds to the maximization of the Newman modularity Q. From the definition of $\tilde{H}_k(z)$ and the equation of state for m it is clear that $m = 0$ for $\gamma \geq 1$. The natural partition of a random graph of any degree distribution with finite variance is the equipartition. This is true for any temperature. In the ground state for $T \to 0$, i.e., $\beta \to \infty$, one further finds from (5.50) that $q \to 1$. The system behaves just as an ordinary SK model with couplings of zero mean. According to (5.56) the ground state energy is obtained from the integral

$$[f] = -\frac{2}{\beta} \sum_k p(g_k) \int_0^\infty Dz \left(\beta J\sqrt{q_k}z + \ln(1 + \exp(-2\beta J\sqrt{q_k}z))\right), \qquad (5.60)$$

where one of the exponentials of the cosh has been factored out and the logarithm has been taken explicitly. The second term in the integral vanishes and one is left with

$$[f] = \sqrt{\frac{2}{\pi}}J \sum_k p(g_k)\sqrt{g_k q}. \qquad (5.61)$$

With $q \to 1$ the ground state energy is finally given by

$$\lim_{\beta \to \infty} [f] = -\sqrt{\frac{2}{\pi}} J \langle g^{1/2} \rangle. \tag{5.62}$$

Though (5.62) is a general result, we are particularly interested in the case $g_i = k_i$, because this corresponds to the modularity Hamiltonian. Let us compare (5.62) to the result of Fu and Anderson, who find from the assumption $p_{ij} = p$ [1]

$$\lim_{\beta \to \infty} [f_{FA}] = -\sqrt{\frac{2}{\pi}} J \sqrt{Np(1-p)}. \tag{5.63}$$

Since $Np \approx \langle k \rangle$, one sees that FA has a ground state energy proportional to the square root of the average degree, while the treatment presented here results in a ground state energy proportional to the average square root of the degree. Since the square root is a concave function, we see that the new approximation results in higher energies for any degree distribution.

The modularity of the ground state partition into two equal-sized parts is then $Q_2 = -[f]/\langle k \rangle$:

$$Q_2 = \sqrt{\frac{2}{\pi}} J \frac{\langle k^{1/2} \rangle}{\langle k \rangle}. \tag{5.64}$$

For comparison, the modularity resulting from FA would be

$$Q_2^{FA} = \sqrt{\frac{2}{\pi}} J \sqrt{\frac{1-p}{\langle k \rangle}}. \tag{5.65}$$

For graphs in which every node has the same degree, the new formulation presented here in (5.64) and the FA result (5.65) coincide. The following numerical experiments will show the adequacy of (5.62) for a number of different degree distributions.

5.2 Numerical Experiments

For comparisons with numerical experiments the k-independent part of the ground state energy of the SK model $U_0 = \sqrt{2/\pi} = 0.798$ [3] was replaced by its replica symmetry breaking counterpart $U_0 = 0.765$ from Ref. [8]. This simply improves the accuracy of our estimates but does not change the qualitative behavior of our predictions.

Random test networks with different degree distributions were created. First, results were checked on Erdős–Rényi (ER) graphs [9,10] with link probability $p_{ij} = p$ and different average degree. Figure 5.1 shows the results of this experiment with networks of $N = 10,000$ nodes and average degree between 3 and 20. The Hamiltonian (5.1) was minimized using simulated annealing [11]. As expected, the ground state was found to have zero magnetization. It is

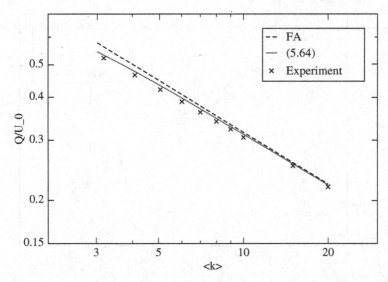

Fig. 5.1. Numerical experiments for ER graphs with $N = 10,000$ nodes and varying average degree $\langle k \rangle$. The correction of the cost function due to optimization Q_2 is given in units of the SK ground state energy U_0. The new formula (5.64) improves the results (5.65) of Fu and Anderson (FA) [1] over the whole range of values and in particular for small average degrees.

remarkable that (5.64) leads to an improved estimate even for ER random graphs. For large $\langle k \rangle$, the two approximations converge as would be expected.

The second ensemble of degree distributions is that of scale-free networks with $N = 10,000$ nodes and a degree distribution of the form $p(k) \propto k^{-\kappa}$. The maximum possible degree was set to 1000 and the graphs were composed using the Molloy–Reed algorithm [12]. The second moment of the distribution exists for all $\kappa \geq 3$. For the experiments $\kappa = 3$ was chosen. In order to produce graphs of different average degree, a minimum degree k_{min} was introduced, such that $p(k < k_{min}) = 0$ with $2 \leq k_{min} \leq 12$. Figure 5.2 shows the result of this experiment. As expected, FA's formula and (5.64) scale identically with the density of the network and approximations are better for denser graphs. Clearly, (5.64) approximates the data points better.

Finally, a second class of scale-free networks is studied. Introducing a k_{min} may have been a too drastic step, as it excludes all nodes of small degree from the network. Therefore, the degree distribution is modified to $p(k) = (k + \Delta k)^{-\kappa}$. Using $\kappa = 3$ as before and varying Δk between 1 and 20, the networks used for the experiments shown in Fig. 5.3 were obtained. Here also the scaling of the cut-size with graph density is different and the improvement of the estimation from (5.64) over FA grows with the average degree in the network.

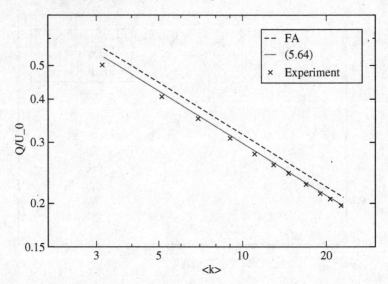

Fig. 5.2. Numerical experiments for scale-free networks with $N = 10,000$ nodes and $p(k) \propto k^{-3}$. Different average degrees were generated by setting $p(k) = 0$ for $k < k_{min}$ with $2 \leq k_{min} \leq 12$. The formula by FA (5.65) [1] underestimates the cut size over the whole range of values and the improvement with (5.64) is roughly constant over the whole range of data points.

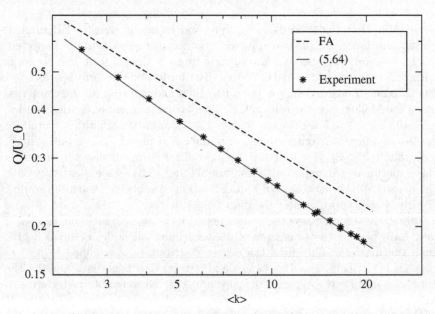

Fig. 5.3. Numerical experiments for scale-free networks with $N = 10,000$ nodes and $p(k) \propto (k + \Delta k)^{-3}$. Different average degrees were generated by varying Δk between $1 \leq \Delta k \leq 20$. Again, the formula by FA (5.65) underestimates the cut size over the whole range of values. The improvement with (5.64) is larger for larger values of $\langle k \rangle$.

In all three cases, the formula by FA underestimates the cut size of the partitioning problem and overestimates the modularity. Hence, (5.64) is proposed as a tighter lower bound on the partitioning problem.

Let us illustrate the contrast between (5.64) and (5.65) in a different way. Both formulas suppose a dependence of the modularity on the degree distribution. If we divide the empirically found modularity by this presumed dependence, we should find straight lines corresponding to the pre-factors in (5.64) or (5.65) only. Figure 5.4 shows the same data points as in Figs. 5.1, 5.2 and 5.3, but rescaled according to (5.65) on the left and according to (5.64) on the right. While using the link density as a parameter as proposed by FA does not lead to a universal curve for the different degree distributions, using the ratio of $\langle k^{1/2} \rangle / \langle k \rangle$ does collapse the data points onto a universal curve. We see that the denser the networks, the better our approximation and that using the replica symmetry breaking value of U_0 improves our accuracy.

Our results from the case of bi-partitioning generalize in a straightforward way to the case of q-partitioning and we can simply replace the scaling from the formulas by Kanter and Sompolinsky (KS) [13] or Lai and Goldschmidt [14]. The expectation value for the maximum modularity of a random graph with arbitrary degree distribution is then [15]

$$Q = 0.97 \frac{\langle \sqrt{k} \rangle}{\langle k \rangle}. \tag{5.66}$$

The factor of $U_0 = 0.97$ corresponds to the ground state energy of the Potts glass in the one-step RSB treatment as calculated by KS [13, 15]. Figure 5.5 shows the maximum modularity obtained when minimizing the Hamiltonian (5.1) in the same graphs as used in Figs. 5.1, 5.2 and 5.3 but with the number of communities q as a free parameter. Again, we see that plotting Q in units

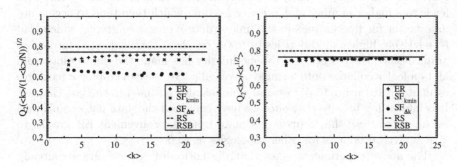

Fig. 5.4. The same data points as in Figs. 5.1, 5.2 and 5.3 for Erdős–Rényi random graphs and two forms of a scale-free (SF) degree distribution, see text for details. *Left*: Scaling Q in units of $\sqrt{(1-p)/Np}$ as suggested by (5.65). *Right*: Scaling Q in units of $\langle \sqrt{k} \rangle / \langle k \rangle$ collapses the data points onto one universal curve as expected from (5.64). *The dashed and solid lines correspond to U_0 in the replica symmetric (RS) and replica symmetry breaking (RSB) case, respectively.*

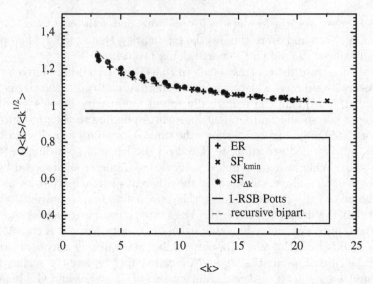

Fig. 5.5. Numerical experiments maximizing the modularity in random graphs of different topologies (Erdős–Rényi (ER) random graphs and two forms of a scale-free (SF) degree distribution, see text for details). Scaling Q in units of $\langle\sqrt{k}\rangle/\langle k\rangle$ collapses the data points onto one universal curve as expected from (5.64). *The solid line* corresponds to ground state energy U_0 of the Potts glass in one-step replica symmetry breaking approximation due to Kanter and Sompolinsky [13]. *The dashed line* corresponds to an estimate obtained from a recursive bi-partitioning along the lines of (4.16) using (5.64).

of $\langle\sqrt{k}\rangle/\langle k\rangle$ collapses the data points onto a single universal curve. The approximation of the universal value of U_0, however, is much slower. This was already observed in the last chapter and can be explained by the fact that sparser graphs tend to cluster into more modules than predicted by KS. This leads to a higher number of degrees of freedom which then tend to accommodate better for fluctuations in the link structure of sparse graphs which lead to relatively higher modularities or lower cut sizes than expected for denser graphs. Since the rescaling of the modularity has made the data points from all topologies collapse onto a single universal curve, we can employ a recursive estimate (4.16) using (5.64) several times as also outlined in the last chapter. It is interesting to note that once we have collapsed the data onto a universal curve, we can use the recursive estimate from the convenient ER graphs to give analytical bounds for other topologies as well.

The above experiments show that the notion of "dense" graphs already applies to those with an average degree of 5 and above. Of course, the results by FA represent a very early result used for comparison in the analysis. It should not go unmentioned that a large amount of work has gone into solving the graph partitioning problem on sparse graphs or in providing replica symmetry breaking solutions. All of this work, however, was focused on ER graphs [16–19] or Bethe lattices [20–24] and resulted in improvements of the approximation as a function of the average connectivity. However, $\langle k \rangle$ is characteristic *only* for graphs with a Poissonian degree distribution or fixed degree.

An approach different from the replica method can also be taken via the cavity method [25–27]. There, however, the solution of self-consistent equations for the local field distribution would be necessary for every degree distribution. This allows for a more accurate estimation of the partitioning cost of graphs with arbitrary degree distribution, but does not result in formulas as handy as those presented here. This method gives good approximations to sparse graphs with loops of length $\mathcal{O}(\ln N)$, i.e., locally tree-like graphs, and hence is complementary to the replica treatment. It will be presented in the following chapter.

References

1. Y. Fu and P. W. Anderson. Application of statistical mechanics to NP-complete problems in combinatorial optimisation. *Journal of Physics A: Mathematical and General*, 19:1605–1620, 1986.
2. H. Nishimori. *Statistical Physics of Spin Glasses and Information Processing*. Oxford Science Publications, Oxford, 2001.
3. D. Sherrington and S. Kirkpatrick. Solvable model of a spin-glass. *Physical Review Letters*, 35(26):1792–1796, 1975.
4. D. -H. Kim, G. J. Rodgers, B. Kahng, , and D. Kim. Spin glass phase-transitions on scale-free networks. *Physical Review E*, 71:056115, 2005.
5. S. F. Edwards and P. W. Anderson. Theory of spin glasses. *Journal of Physics F: Metal Physics*, 5:965, 1975.
6. J. Hubbard. Calculation of partition functions. *Physical Review Letters*, 3:77–78, 1959.
7. A. Engel and C. Van den Broeck. *Statistical Mechanics of Learning*. Cambridge University Press, New York, 2001.
8. G. Parisi. A sequence of approximated solutions to the s-k model for spin glasses. *Journal of Physics A: Mathematical and General*, 13:L115–L121, 1980.
9. P. Erdős and A. Rényi. On the evolution of random graphs. *Publications of the Mathematical Institute of the Hungarian Academy of Sciences*, 5:17–61, 1960.
10. B. Bollobas. *Random Graphs*. Cambridge University Press, New York, 2nd edition, 2001.
11. S. Kirkpatrick, C.D. Gelatt Jr., and M.P. Vecchi. Optimization by simulated annealing. *Science*, 220:671–680, 1983.
12. M. Molloy and B. Reed. A critical point for random graphs with given degree sequence. *Random Structures and Algorithms*, 6:161–179, 1995.
13. I. Kanter and H. Sompolinsky. Graph optimisation problems and the Potts glass. *Journal of Physics A: Mathematical and General*, 20:L636–679, 1987.
14. P. -Y. Lai and Y. Y. Goldschmidt. Application of statistical mechanics to combinatorial optimization problems: The chromatic number problem and qpartitioning of a graph. *Journal of Statistical Physics*, 48(3/4):513–529, 1987.
15. J. Reichardt and S. Bornholdt. Statistical mechanics of community detection. *Physical Review E*, 74:016110, 2006.

16. W. Liao. The graph-bipartitioning problem. *Physical Review Letters*, 59(15):1625–1628, 1987.

17. W. Liao. Replica-symmetric solution of the graph-bipartitioing problem. *Physical Review A*, 37:587–595, 1988.

18. Y. Y. Goldschmidt and C. De Dominicis. Replica symmetry breaking in the spin-glass model on lattices with finite connectivity: Application to graph partitioning. *Physical Review B*, 410(4):2184–2197, 1990.

19. I. Kanter and H. Sompolinsky. Mean-field theory of spin-glasses with finite coordination number. *Physical Review Letters*, 58(2):164–167, 1987.

20. K. Y. M. Wong and D. Sherrington. Graph bipartitioning and spin glasses on a random network of finite valence. *Journal of Physics A: Mathematical and General*, 20:L793–L799, 1987.

21. M. Mezard and G. Parisi. Mean-field theory of randomly frustrated systems with finite connectivity. *Europhysics Letters*, 3(10):1067–1074, 1987.

22. J. R. Banavar, D. Sherrington, and N. Sourlas. Graph bipartioning and statistical mechanics. *Journal of Physics A: Mathematical and General*, 20:L1–L8, 1987.

23. M. J. de Oliveira. Graph optimization problems on the bethe lattice. *Journal of Statistical Physics*, 54(1/2):477–493, 1989.

24. D. Sherrington and K. Y. M. Wong. Graph bipartitioning and the bethe spin glass. *Journal of Physics A: Mathematical and General*, 20:L785–L791, 1987.

25. M. Mezard and G. Parisi. The cavity method at zero temperature. *Journal of Statistical Physics*, 111(1/2):1–34, 2003.

26. O. Martin, R. Monasson, and R. Zecchina. Statistical mechanics methods and phase transitions in optimization problems. *Theoretical Computer Science*, 256:3–67, 2001.

27. S. Franz, M. Leone, and F. L. Toninelli. Replica bounds for diluted nonpoissonian spin systems. *Journal of Physics A: Mathematical and General*, 36:10967–10985, 2003.

6

Modularity of Sparse Random Graphs

The connection between community detection and graph partitioning has been shown in the last chapters as well as the importance of giving expectation values of the modularity or cut size for partitions of random graphs. However, all the approximation formulas given up to this point have assumed dense graphs for which the distribution of local fields in the ground state can be approximated by a Gaussian. In this chapter, this approximation will be dropped and the distribution of local fields will be calculated explicitly for any degree distribution.

In the last chapter, the problem of graph partitioning was interpreted as finding a ground state of an infinite range spin glass with a coupling distribution of zero mean. It can also be interpreted as finding the ground state of a ferromagnetic q-state Potts model [1] under the constraint of zero magnetization. This is the approach taken in this chapter. Note that now the system is a sparse system. Couplings exist only between connected spins. Further, it is assumed that the loops in the graph are long ($\mathcal{O}(\ln N)$), i.e., there exist hardly any triangles, etc. This is the case for any sparse random graph in the thermodynamic limit. Such graphs are called locally tree like.

A number of results exist for graphs with fixed connectivity [2–7]. Here a statistical mechanics approach to the problem of q-partitioning a locally tree-like graph is presented that can deal with arbitrary degree distributions and also allows for correlations of the degrees of neighboring vertices.

6.1 Graph Partitioning Using the Cavity Method

The statistical mechanics formulation of the q-partitioning problem is done via the following ferromagnetic Potts Hamiltonian:

$$\mathcal{H}_F(\{\sigma\}) = - \sum_{i \neq j} J_{ij}\delta(\sigma_i, \sigma_j), \qquad (6.1)$$

Reichardt, J.: *Modularity of Sparse Random Graphs*. Lect. Notes Phys. **766**, 87–118 (2009)
DOI 10.1007/978-3-540-87833-9_6 © Springer-Verlag Berlin Heidelberg 2009

where J_{ij} is the $\{0,1\}$ adjacency matrix of the graph and σ_i denotes the Potts spin variable with $\sigma_i \in \{1,2,...,q\}$. Once one finds the ground state under the constraint $\sum_i \delta(\sigma_i, \tau) = N/q$ for all $\tau \in \{1,2,...,q\}$, one can write the total number of cut edges C in the system using the ground state energy E_g of the above Hamiltonian (6.1):

$$C_q = M + E_g = M\left(\frac{q-1}{q} - Q_q\right). \tag{6.2}$$

Note the difference to (5.5). Also note that the modularity of the q-partition Q_q can be expressed via Hamiltonian (6.1) as

$$Q_q = -\frac{\mathcal{H}_F}{M} - \frac{1}{q}. \tag{6.3}$$

This expression is only valid for magnetization zero, i.e., an exact q-partition.

6.1.1 Cavity Method at Zero Temperature

The ground state energy of (6.1) can be calculated by applying the cavity method at zero temperature following the approach presented by Mezard and Parisi [8] in the formulation for a Potts model as presented by Braunstein et al. [9,10] for coloring random graphs. The energy of a system of N spins is written as dependent on a "cavity spin" σ_1 via the "cavity field" $\boldsymbol{h}_1$:

$$E^N(\sigma_1) = A - \sum_{\tau=1}^{q} h_1^\tau \delta(\tau, \sigma_1). \tag{6.4}$$

Note that h_1^τ takes only integer values, if J_{ij} is composed of only $\{0,1\}$. The components of the cavity field $\boldsymbol{h}_i$ denote the change in energy of the system with a change in spin i. In general, these are different from the "effective fields" $\sum_j J_{ij}\sigma_j$ acting on spin σ_i, which are used to calculate the magnetization. Adding a new spin σ_0 connected to σ_1, the energy of the now $N+1$ spin system is a function of both σ_1 and σ_0:

$$E^{N+1}(\sigma_1, \sigma_0) = A - \sum_{\tau=1}^{q} h_1^\tau \delta(\tau, \sigma_1) - J_{10}\delta(\sigma_1, \sigma_0). \tag{6.5}$$

One can now write this expression in such a way that it only depends on the newly added cavity spin σ_0:

$$E^{N+1}(\sigma_0) = \min_{\sigma_1} E^{N+1}(\sigma_1, \sigma_0) \equiv A - w(\boldsymbol{h}_1) - \sum_{\tau=1}^{q} \hat{u}^\tau(J_{10}, \boldsymbol{h}_1)\delta(\tau, \sigma_0). \tag{6.6}$$

The functions w and $\hat{u}$ take the following form:

$$w(\boldsymbol{h}) = \max(h^1, ..., h^q), \tag{6.7}$$
$$\hat{u}^\tau(J, \boldsymbol{h}) = \max(h^1, ..., h^\tau + J, ..., h^q) - w(\boldsymbol{h}). \tag{6.8}$$

From (6.8) one sees that $\hat{u}^\tau(\boldsymbol{h})$ is one, whenever the τth component of $\boldsymbol{h}$ is maximal with respect to all other components in $\boldsymbol{h}$ and zero otherwise. Due to possible degeneracy in the components of $\boldsymbol{h}$, the vector $\hat{u}(\boldsymbol{h})$ may have more than one non-zero entry and is never completely zero.

One can interpret this "cavity bias" $\hat{u}(J_{10}, \boldsymbol{h_1})$ as a new cavity field $\boldsymbol{h_0}$ which describes the change of the energy of the system under a change of σ_0. The field $\boldsymbol{h_1}$ which was formerly acting only on σ_1 has now been propagated to σ_0 and acts there as cavity field $\boldsymbol{h_0} = \hat{u}(J_{10}, \boldsymbol{h_1})$. In general cases where the new spin σ_0 is connected to d different cavity spins, the cavity biases have to be combined linearly to give the cavity field $\boldsymbol{h_0} = \sum_{i=1}^{d} \hat{u}(J_{i0}, \boldsymbol{h_i})$.

With this, one "iteration" has just been completed and the basic idea behind the Bethe–Peierls approach [11], which is the foundation of the cavity method, has been demonstrated. The goal is to find a distribution of the cavity fields $P_{\mathrm{cav}}(\boldsymbol{h})$ which is stable under this iteration procedure and site independent. For trees this is granted, and for graphs which are locally tree like, i.e., without short loops, this is at least approximately true and corresponds to the assumption of a replica symmetric ground state. It turns out that the entire problem of finding the ground state properties of the system is reduced to the existence and actual finding of the distribution of the cavity fields.

The iteration procedure can also be interpreted as a form of message passing. The spins σ_i see cavity fields $\boldsymbol{h_i}$ in the absence of spin σ_0 and send "messages" $\hat{u}(J_{i0}, \boldsymbol{h_i})$ along the link J_{i0} to spin σ_0. Spin σ_0 collects these messages to form a cavity field which is then passed to some other node j in the form of a message. The cavity field is hence a field that a node i sees in the absence of node j. It is transformed into a message and passed from node i to j. A node of degree k sees k different cavity fields, each made from $k - 1$ messages. There are hence twice as many cavity fields and messages as there are links in the graph. The cavity fields and messages "live" on the edges of the graph.

At this point let us recall the definition of the excess degree of a node from Sect. 1.2 and Ref. [12]. The excess degree d is nothing but the number of links a node i has minus one: $d_i = k_i - 1$ and hence the number of messages or inputs that are used in the calculation of the k_i cavity fields of that node. Since the cavity fields and biases live on the edges of the graph, drawing from their distribution and averaging over their distribution means drawing from and averaging over the set of edges.

The number of messages that are used in the calculation of the cavity field for a particular edge is distributed as $q(d)$, the probability of finding a node of excess degree d by following a randomly chosen link. Hence, it must satisfy $q(d) \propto (d+1)p(d+1)$, the degree of the node $k = d + 1$ times the probability of drawing a node of degree k at random from the set of *nodes*. Correctly normalized one has

$$q(d) = \frac{(d+1)p(d+1)}{\sum_k p(k)k} = \frac{(d+1)p(d+1)}{\langle k \rangle}. \tag{6.9}$$

Only for the case of a Poissonian degree distribution $p(k) = e^{-\langle k \rangle}\frac{\langle k \rangle^k}{k!}$ one finds $q(d) = p(d)$, i.e., the excess degree is distributed in the same way as the degrees themselves.

With these definitions, one can write self-consistent equations for the probability distributions of the cavity fields $P_{\mathrm{cav}}(h)$ and the cavity biases $\mathfrak{Q}(u)$. Do not confuse the distribution of messages $\mathfrak{Q}$ and the modularity Q. Note, how $q(d)$, the distribution of the excess degree, enters into the equation. Furthermore, one can write equations for the distribution of effective fields $P_{\mathrm{eff}}(h)$ that "live" on the nodes of the graph. This distribution is needed later in the calculation of the ground state energy. Note how $p(k)$, the degree distribution, enters into its calculation:

$$P_{\mathrm{cav}}(h) = \sum_{d=0}^{\infty} q(d) \int \prod_{i=1}^{d} (d^q h_i P_{\mathrm{cav}}(h_i))\, \delta\left(h - \sum_{i=1}^{d} \hat{u}(h_i)\right), \tag{6.10}$$

$$\mathfrak{Q}(u) = \int d^q h\, P_{\mathrm{cav}}(h)\delta(u - \hat{u}(h)), \tag{6.11}$$

$$\mathfrak{Q}(u) = \sum_{d=0}^{\infty} q(d) \int \prod_{i=1}^{d} (d^q u_i \mathfrak{Q}(u_i))\, \delta\left(u - \hat{u}\left(\sum_{i=1}^{d} u_i\right)\right), \tag{6.12}$$

$$P_{\mathrm{eff}}(h) = \sum_{k=0}^{\infty} p(k) \int \prod_{i=1}^{k} (d^q h_i P_{\mathrm{cav}}(h_i))\, \delta\left(h - \sum_{i=1}^{k} \hat{u}(h_i)\right), \tag{6.13}$$

$$P_{\mathrm{eff}}(h) = \sum_{k=0}^{\infty} p(k) \int \prod_{i=1}^{k} (d^q u_i \mathfrak{Q}(u_i))\, \delta\left(h - \hat{u}\left(\sum_{i=1}^{k} u_i\right)\right). \tag{6.14}$$

As a special case, the distinction between effective fields and cavity fields is irrelevant for graphs with Poissonian degree distribution because of $q(d) = p(d)$. Note that these equations can be solved via an iteration procedure. Plugging a test function, e.g., $P_{\mathrm{cav}}(h)$, into the right-hand side of (6.10) one obtains a new $P_{\mathrm{cav}}(h)$ after integration and summation. This process is repeated until a fix-point distribution $P_{\mathrm{cav}}(h)$ is reached which is a solution of (6.10).

The energy of the system is then calculated from the contribution of a site addition ΔE_1, i.e., adding a vertex of degree k to the graph, and the contribution of a link removal ΔE_2 as outlined in Ref. [8]. The change in energy due to a site addition is

$$\Delta E_1 = -\sum_{k=0}^{\infty} p(k) \int \prod_{i=1}^{k} (d^q u_i \mathfrak{Q}(u_i))w\left(\sum_{i=1}^{k} u_i\right) = -\int d^q h\, P_{\mathrm{eff}}(h)w\,(h) \tag{6.15}$$

Note that the site addition is different from the iteration procedure, as now the full degree of the vertex and the effective field plays a role. The energy of a link addition (negative link removal) is

$$\Delta E_2 = \int d^q \boldsymbol{h}_1 d^q \boldsymbol{h}_2 P_{\text{cav}}(\boldsymbol{h}_1) P_{\text{cav}}(\boldsymbol{h}_2) \left(w(\boldsymbol{h}_1) - w(\boldsymbol{h}_1 + \hat{u}(\boldsymbol{h}_2)) \right). \qquad (6.16)$$

Here, the cavity fields at the ends of the edge to remove play a role. The energy density of the system is then written as

$$E = \Delta E_1 - \frac{\langle k \rangle}{2} \Delta E_2. \qquad (6.17)$$

There is a subtle point here about (6.17). It is necessary to keep the degree distribution $p(k)$ invariant when going from N to $N+1$ sites by adding a new site with k links drawn from $p(k)$. With the k links from the new node, one increases the total number of stubs (ends of edges) of the N old nodes by k, therefore changing the degree distribution of the old nodes, which already is $p(k)$ and should be kept invariant. Hence, one needs to cut $k/2$ links among the old N nodes in order to create the k stubs to which the new node can be connected *without* changing the degree distribution. Averaged over $p(k)$, one needs to cut $\langle k \rangle / 2$ links before adding the new node as in (6.17). Note that these expressions are completely general and do not depend on the specific Hamiltonian under study as this is encoded in the specific form of $\hat{u}$ and w [8].

6.1.2 Symmetry Conditions

The description of the system in terms of cavity fields and biases is equivalent and we have chosen a formulation in terms of biases or messages only. We have already seen from (6.8) that the possible messages are the corners of a hypercube in q dimensions except the corner at $\boldsymbol{0}$. Therefore, there are in principle $2^q - 1$ different possible messages. Since one wants to find solutions of an equipartitioning problem, one is only interested in solutions which are symmetric under an arbitrary permutation of the spin indices, i.e., the solution must be fully color symmetric. An *ansatz* that bears this symmetry is

$$\mathcal{Q}(\boldsymbol{u}) = \eta_\tau \text{ with } \tau = \|\boldsymbol{u}\|^2. \qquad (6.18)$$

This means that one is simply counting the number of ones in a message and this number is represented by the index of the order parameter η_τ. All messages with the same number of ones are equally probable. Hence, one only needs to determine q different probabilities η_τ, of which $q-1$ are independent due to the normalization constraint:

$$\sum_{\tau=1}^{q} \binom{q}{\tau} \eta_\tau = 1. \qquad (6.19)$$

The paramagnetic solution $\eta_q = 1$ and $\eta_{\tau \neq q} = 0$ is always possible but shall not be taken into consideration as it is unstable.

With these considerations only, we can already draw a conclusion on the dependency of ΔE_2 on the order parameters. From (6.16) we see that the integrant is only non-zero whenever $\boldsymbol{u}_1$ and $\boldsymbol{u}_2$ have at least one non-zero entry in common. Hence ΔE_2 is the negative of one minus the probability that $\boldsymbol{u}_1$ and $\boldsymbol{u}_2$ with τ_1 and τ_2 non-zero entries, respectively, do not have an overlapping one:

$$\Delta E_2 = \sum_{\tau_1=1}^{q} \sum_{\tau_2=1}^{q-\tau_1} \frac{q!}{\tau_1! \tau_2! (q - \tau_1 - \tau_2)!} \eta_{\tau_1} \eta_{\tau_2} - 1. \qquad (6.20)$$

This expression may also be read as the negative of the probability that two nodes may "agree" on a common spin state to satisfy the link between them. The above expression holds for any distribution of excess degrees.

As an example, let us study a network with fixed connectivity, a Bethe lattice with $k = 3$ and $q = 2$, i.e., a bi-partition. The excess degree of every node is two. The self-consistent equation for the messages can be cast into a system of non-linear polynomial equations for the $q = 2$ order parameters:

$$\eta_1 = \eta_1^2 + 2\eta_1 \eta_2, \qquad (6.21)$$
$$\eta_2 = 2\eta_1^2 + \eta_2^2, \qquad (6.22)$$
$$1 = 2\eta_1 + \eta_2. \qquad (6.23)$$

It is instructive to interpret this system of equations. Every node has two "inputs" over which it can receive messages and one "output". The first equation means a node sees a non-degenerate maximum in the cavity field with probability η_1, because it either has two also non-degenerate inputs pointing in the same direction (this happens with probability η_1^2) or it has one non-degenerate input and one twofold-degenerate input (which happens with probability $2\eta_1 \eta_2$). The second equation means a node sees a twofold-degenerate cavity field with probability η_2, because it has two non-degenerate inputs pointing in different directions (this happens with probability $2\eta_1^2$) or it has two twofold-degenerate inputs (which happens with probability η_2^2). The third equation is simply the normalization condition. The solution of this system is given by $\eta_1 = 1/3$ and $\eta_2 = 1/3$. Formally, these equations are equivalent to those derived for an Ising spin glass with couplings $J_{ij} \pm 1$ on a Bethe lattice and the results can be applied immediately [8].

6.1.3 Bi-partitioning

Due to the large number of combinations of messages for large k and the large number of different messages for large values of q, it was not possible to find a simple analytic expression for the coefficients in the self-consistent calculation of the order parameters $\eta_1 ... \eta_q$ for arbitrary k and q. The expression is simple though, if only two spin states are allowed. The probability for a cavity field $\boldsymbol{h} = (h_1, h_2)$ seen by a node of excess degree d is expressed as

$$P_{\text{cav}}^d(h_1, h_2) = \frac{d!}{(d - h_1)!(d - h_2)!(h_1 + h_2 - d)!} \eta_1^{2d - h_1 - h_2}(1 - 2\eta_1)^{h_1 + h_2 - d}.$$

(6.24)

The average over the excess degree distribution then reads

$$P_{\text{cav}}(h_1, h_2) = \sum_{d=0}^{\infty} q(d) P_{\text{cav}}^d(h_1, h_2).$$

(6.25)

Recall that η_τ is the probability that the maximum component of the cavity field is τ-fold degenerate. One then has

$$\eta_1 = \sum_{h_1=1}^{\infty} \sum_{h_2=0}^{h_1-1} P_{\text{cav}}(\boldsymbol{h} = (h_1, h_2)),$$

(6.26)

$$\eta_2 = 1 - 2\eta_1 = \sum_{h=1}^{\infty} P_{\text{cav}}(\boldsymbol{h} = (h, h)),$$

(6.27)

which can also be understood as a self-consistent equation for the order parameters η_τ and can be solved easily in an iterative manner again. For a partition into only two parts, we only need to determine two order parameters η_1 and η_2. The normalization condition (6.19) reduces the problem to determining only a single order parameter as $\eta_2 = 1 - 2\eta_1$. We formulate the problem in terms of η_2:

$$\eta_2 = \sum_{n_0=0}^{\infty} \sum_{n=0}^{\infty} q(n_0 + 2n) \frac{(n_0 + 2n)!}{n_0! n! n!} \left(\frac{1 - \eta_2}{2}\right)^{2n} \eta_2^{n_0}$$

(6.28)

$$= \sum_{d=0}^{\infty} q(d) \sum_{n=0}^{\lfloor \frac{d}{2} \rfloor} \frac{d!}{(d - 2n)! n! n!} \left(\frac{1 - \eta_2}{2}\right)^{2n} \eta_2^{d - 2n}.$$

(6.29)

These equations can be easily iterated for any excess degree distribution $q(d)$ to find the order parameters of a bi-partition to arbitrary accuracy. In case of a Poissonian distribution $q(d) = e^{-\lambda} \lambda^d / d!$ with mean λ, the two sums in (6.28) decouple and we find

$$\eta_2 = e^{-\lambda(1 - \eta_2)} I_1(0, \lambda(1 - \eta_2)),$$

(6.30)

where $I_1(0, x)$ is the modified Bessel function of the first kind. The energy per link ΔE_2 is given by $\Delta E_2 = 2\eta_1^2 - 1$ according to (6.20).

Figure 6.1 shows the order parameter η_1 and the ground state energy for Bethe lattices and ER random graphs with different connectivities. Note how the values of η_1 for Bethe lattices with even connectivity (odd excess degree) always lie above the curve for ER random graphs, while those with

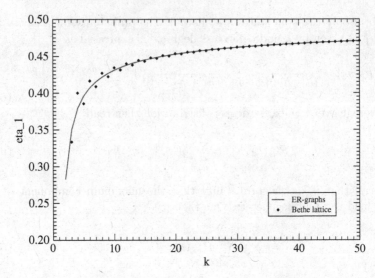

Fig. 6.1. The order parameter η_1 for ER random graphs and Bethe lattices vs. the average connectivity.

odd connectivity (even excess degree) lie below them. This is understandable from the fact that odd excess degrees favor an asymmetry in the cavity fields and hence favor η_1. The difference between Bethe lattices and ER random graphs becomes negligible for larger degrees.

In order to find the cut size we need an expression for the ground state energy. The energy per nodes is given as

$$\Delta E_1 = -\sum_{n_0=0}^{\infty}\sum_{n_1=0}^{\infty}\sum_{n_2=0}^{\infty} p(n_0+n_1+n_2)\frac{(n_0+n_1+n_2)!}{n_0!n_1!n_2!}\eta_0^{n_0}\eta_1^{n_1+n_2}[n_0+\max(n_1,n_2)].$$

(6.31)

For any degree distribution $p(k)$, this can be rewritten as

$$\Delta E_1 = -\langle k\rangle(\eta_2 + 2\eta_1(\eta_1+\eta_2) + X) = -\langle k\rangle(1 - 2\eta_1^2 + X) = -\langle k\rangle(X - \Delta E_2),$$

(6.32)

where we have used the implicit equations for the order parameters η_1 and η_2 from (6.28) and introduced a function X, which is defined as

$$X = \frac{1}{\langle k\rangle}\sum_{n_0=0}^{\infty}\sum_{n=0}^{\infty} p(n_0+2n)\frac{(n_0+2n)!}{n_0!n!n!}\eta_1^{2n}\eta_2^{n_0}n$$

(6.33)

$$= \frac{1}{\langle k\rangle}\sum_{k=2}^{\infty} p(k)\sum_{n=0}^{\lfloor\frac{k}{2}\rfloor}\frac{k!}{(k-2n)!n!n!}\eta_1^{2n}\eta_2^{k-2n}n.$$

(6.34)

In case of a Poissonian degree distribution $p(k)$ with mean λ we can write for X even shorter

$$X_\lambda = \eta_1 e^{-2\lambda\eta_1} I_1(1, 2\lambda\eta_1).$$

(6.35)

Finally, we can write for the ground state energy per node (6.17) and modularity Q_2 (6.3) of the bi-partition the following expressions:

$$E = -\frac{\langle k \rangle}{2} \left(1 + 2(X - \eta_1^2) \right),$$ (6.36)

$$Q_2 = \frac{1}{2} + 2(X - \eta_1^2).$$ (6.37)

We thus have a formulation which depends entirely on the order parameters. Since the energy per node can never exceed $-\langle k \rangle/2$ and η_1 is bounded by $\eta_1 < 1/2$ we have a bound on X which is $X \leq 1/4$. With these expressions we can find expected cut sizes for the bi-partitioning problem and expected modularities for random networks of arbitrary degree distributions.

Let us compare these results to numerical experiments. Banavar et al. [13] have performed numerical studies of the graph bi-partitioning problem for Bethe lattices. For lattices of varying degree, Table 6.1 compares the results of the cavity method Q_2 and the approximation formula (5.64) which is in the case of fixed connectivity equivalent to the approximation by Fu and Anderson (5.65) Q_2^{FA}, with the numerical results by Banavar et al. Q_{Banavar} [13]. For the case of $d+1 = 3$, the problem has also been studied by Wong and Sherrington [3,6], Mezard and Parisi [4] and De Oliveira [14].

One can observe that the results of the cavity method give an improved estimate over the replica results (5.64) or (5.65) for sparse graphs. The agreement with the numerical results is within a few percent and can be improved by resorting to the replica symmetry breaking formalism [8]. For denser graphs, the finite size effects of the numerical simulations become more pronounced. The largest network studied by Banavar et al. had only 4000 nodes. Since the cavity method relies on the absence of short loops the approximation will inevitably break down. This explains why for denser graphs the replica results give the better approximation, the networks studied by Banavar simply are not exactly tree like for large connectivities.

Table 6.1. Comparison of the numerical results for bi-partitioning random Bethe lattices of varying connectivity $d + 1 = k$ by Banavar et al. Q_{Banavar} [13] and the results of the cavity method Q_2 and those of the replica method by Fu and Anderson (5.65) Q_2^{FA} [15].

$d+1$	η_1	E_g	Q_2	Q_2^{FA}	Q_{Banavar}
3	0.333333	−1.388889	0.426	0.441	0.420
4	0.4	−1.744	0.372	0.382	0.366
5	0.385604	−2.095665	0.338	0.341	0.332
6	0.416453	−2.427626	0.309	0.312	0.303
8	0.426902	−3.081101	0.270	0.270	0.264
9	0.422281	−3.402895	0.256	0.254	0.250
10	0.434291	−3.715701	0.243	0.241	0.235
15	0.442999	−5.252232	0.200	0.197	0.193
20	0.453388	−6.741082	0.174	0.171	0.168

The previous chapter suggested via a replica calculation that the modularity of random networks scales universally as $\langle \sqrt{k} \rangle / \langle k \rangle$. Let us investigate if this holds true for the results obtained by the cavity method too. Figure 6.2 shows modularities Q_2 of bi-partitions for networks with the same degree distributions as used in the previous chapter. In particular, we compare the results for the Poissonian degree distribution of Erdös–Renyi (ER) graphs [16] with mean λ, i.e., $p(k) = e^{-\lambda} \lambda^k / k!$, with two types of scale-free degree distributions. The scale free distributions are of the stretched form $p(k) \propto (k + \Delta k)^{-\gamma}$ (SF Δk) and of the standard form $p(k) \propto k^{-\gamma}$ (SF k_{min}) but with a minimum degree k_{min} such that $p(k < k_{min}) = 0$. We varied Δk between 1 and 50 and k_{min} between 2 and 30 while using $\gamma = 3$ in both distributions. Early replica calculations for ER graphs [15] had predicted a scaling of the modularity as $Q_2 = U_0 \sqrt{(1-p)/\langle k \rangle}$ which for $p \ll 1$ can be interpreted as scaling with $\langle k \rangle^{-1/2}$. As the left part of Fig. 6.2 shows, this scaling does not hold universally for the scale-free degree distributions.

The right part of Fig. 6.2 shows that the data points converge to the same value when plotting Q_2 in units of $\langle \sqrt{k} \rangle / \langle k \rangle$. Thus, the cavity calculation in this chapter recovers the universal dependence of Q_2 on the two moments $\langle \sqrt{k} \rangle$ and $\langle k \rangle$ as found in the last chapter.

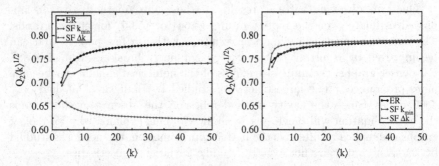

Fig. 6.2. The modularity of bi-partition Q_2 in random graphs with different topologies. We studied graphs with a Poissonian degree distribution $p(k) = e^{-\lambda} \lambda^k / k!$ (ER) and two types of scale-free degree distributions. The first one is a stretched power law (SF Δk) and has the form $p(k) = (k + \Delta k)^{-\gamma}$ with $\Delta k \in [17, 18]$, while the second (SF k_{min}) has the form $p(k) = k^{-\gamma}$, but with a varying minimum degree k_{min} with $k_{min} \in [19, 20]$. For both scale-free distributions we choose $\gamma = 3$. *Left:* We plot the modularity Q_2 in units of $\langle k \rangle^{-1/2}$ as suggested by earlier replica calculations (5.65) [15] and find the results to depend on the form of the degree distribution. *Right:* We plot the modularity Q_2 in units of $\langle \sqrt{k} \rangle / \langle k \rangle$ as suggested by (5.64) and find that this choice collapses the data onto a universal line in the limit of dense graphs.

6.1.4 Limit of Dense Graphs with Poissonian Degree Distribution

The special form of the Poissonian degree distribution has led to numerous simplifications so far. Two crucial simplifications are the cancellations of the term $(n_0 + 2n)!$ which decouples the sums in (6.28) and the fact that for this distribution the degree and excess degree distributions are indeed the same which simplifies the calculation of the energy per node in (6.33).

These simplifications allow us to investigate the scaling of the ground state energies of the bi-partitioning problem for Poissonian graphs in the limit of large average degree. We will show that this allows us to recover the results of the replica calculations of Fu and Anderson [15] plus correction terms. The Bessel functions in (6.28) can be approximated for large arguments $x \gg n$ and fixed η as

$$I_1(n, x) \approx \frac{e^x}{\sqrt{2\pi x}}. \tag{6.38}$$

Using this approximation we obtain for the order parameter η_1 the following equation:

$$2\eta_1 \approx 1 - (4\pi\lambda\eta_1)^{-1/2}. \tag{6.39}$$

Equation (6.35) is approximated using (6.38) and (6.39) as

$$X_\lambda \approx \frac{1}{2}\sqrt{\frac{1}{\pi\lambda}}\sqrt{\eta_1} = \eta_1 - 2\eta_1^2. \tag{6.40}$$

Now we expand the solution of (6.39) in powers of $1/\lambda$ which leads to an approximation for η_1 and hence η_1^2 as

$$\eta_1 \approx \frac{1}{2} - \frac{1}{4}\sqrt{\frac{2}{\pi\lambda}} - \frac{1}{8\pi\lambda} + \mathcal{O}(\lambda^{-3/2}) \text{ and} \tag{6.41}$$

$$\eta_1^2 \approx \frac{1}{4} - \frac{1}{4}\sqrt{\frac{2}{\pi\lambda}} + \mathcal{O}(\lambda^{-3/2}). \tag{6.42}$$

This shows how η_1 approaches the limit of $1/2$ as λ goes to infinity as expected. Pluggings (6.40), (6.41) and (6.42) into (6.36) leads to

$$Q_2 \approx \frac{1}{2} + 2(\eta_1 - 3\eta_1^2) \approx \sqrt{\frac{2}{\pi\lambda}} - \frac{1}{4\pi\lambda} + \mathcal{O}(\lambda^{-3/2}), \tag{6.43}$$

which is to be compared to the result from the replica calculation of Fu and Anderson for dense graphs with connection probability p which yielded $Q_2 = \sqrt{2/\pi}\sqrt{(1-p)/(pN)}$. For small p and identifying $\lambda = pN$ we see that the cavity method brings a correction term of the order of $1/\lambda$ to the results of the replica method. We can interpret this result also in terms of whether a graph is dense or not. As we see, applying a result obtained for a dense graph gives a result which differs in the order of $1/\langle k \rangle$ from the result obtained when considering a sparse graph.

6.1.5 q-Partitioning of a Bethe Lattice with three Links per Node

Thus far we have dealt with bi-partitions of graphs with arbitrary degree distribution as one of the cases where we can write the field equations as a system of coupled polynomials. The other special case for which this can be done is a Bethe lattice where every node has exactly $k = 3$ neighbors. Then, every edge leads to a node with excess degree $d = 2$ and we can write for the order parameters η_τ the following equation:

$$\eta_\tau = \sum_{\alpha=1}^{\tau-1} \binom{\tau}{\alpha} \eta_\alpha \eta_{\tau-\alpha} + \eta_\tau^2 + 2\eta_\tau \sum_{\alpha=1}^{q-\tau} \binom{q-\tau}{\alpha} \eta_{\tau+\alpha}$$
$$+ \sum_{\alpha=1}^{q-\tau} \sum_{\beta=1}^{q-\tau-\alpha} \binom{q-\tau}{\alpha} \binom{q-\tau-\alpha}{\beta} \eta_{\tau+\alpha} \eta_{\tau+\beta}. \tag{6.44}$$

This is easily interpreted. A message with τ non-zero entries can be formed by combining two messages, one with $\alpha < \tau$ and one with $\tau - \alpha$ non-zero entries which do not overlap as in the first term. Then, two messages with exactly τ non-zero entries may overlap as in the second term. The third term denotes the possibility of combining one message with τ and one with $\alpha > \tau$ non-zero entries, while the last stands for the possibility of having an overlap of exactly τ non-zero entries when combining two messages which both have more than τ non-zero entries.

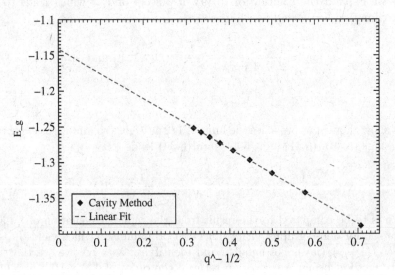

Fig. 6.3. Ground state energy per spin of the q-state ferromagnetic Potts Hamiltonian under the constraint of zero magnetization for Bethe lattices of connectivity $d + 1 = 3$. Data points are taken from Table 6.2.

With the order parameters at hand, we can write the energy per link directly using (6.20). For the energy per node, unfortunately, we cannot write a simple expression for all numbers of parts q and have to calculate ΔE_1 by using Monte Carlo methods. It is interesting to study the ground state energy and modularity as a function of the number of parts q. Naturally, the absolute value of the ground state energy decreases as we divide the random lattice into more and more parts. However, when looking at the modularity, we see that the term $1/q$ which we have to subtract from the negative value of the energy in (6.3) decreases for larger numbers of Q. Plotting E_{gs} vs. $q^{-1/2}$ in Fig. 6.3 we observe a linear dependence which together with (6.3) suggests the existence of an optimal number of q which maximizes the modularity Q_q. Empirically, we find by fitting our data

$$E_g(q) = E_\infty - \frac{B}{\sqrt{q}}, \tag{6.45}$$

with $E_\infty = -1.141$ and $B = 0.3496$. This is a remarkable result as it shows that even for large numbers of q we can still satisfy 2.3 of the 3 connections per node on average. This is not much less than the 2.78 links per node which can be satisfied when partitioning in only two parts. This also means that practically every node has two or more links into its own community, which again means that every random Bethe lattice of connectivity $d + 1 = 3$ has a community structure if the definitions of Radicchi et al. are applied. Plugging (6.45) into (6.3) we can find the number of parts which maximizes Q_q as

$$q^* = \frac{k^2}{B^2} \approx 74. \tag{6.46}$$

The dependence of E_g on q as shown here thus leads to an expectation value for the number of clusters in a random network.

Finally, let us compare these results to replica calculations and numerical experiments of maximizing the modularity by simulated annealing for Bethe lattices of size $N = 10,000$. Table 6.2 summarizes the results together with the order parameters obtained for each value of q. Clearly, the results from the cavity method give a good approximation of the modularity for all numbers of parts q. The replica treatment of KS in (4.13) exhibits a different behavior. Regardless of the connectivity of the network, maximum modularity is achieved for $q = 5$ and modularity decreases with increasing q. The cavity method, however, suggests that for sparse graphs, modularity increases until $q \approx 74$ for Bethe lattices of connectivity $d + 1 = 3$ and decreases only then. The numerical experiments of Chap. 4 (compare Fig. 4.11) have shown that this behavior generalizes to other kinds of sparse graphs as well. The sparser a graph, the higher the value of q^* for which maximum modularity is achieved. With increasing density of the graph, q^* decreases to the replica value of $q_R^* = 5$. The advantage of the cavity method is that it can capture this behavior in a single formalism without the introduction of recursive bipartitionings as introduced in Chap. 4.

Table 6.2. Order parameters and ground state energies of a Bethe lattice with $d + 1 = 3$ links per vertex and different numbers of spin states q. Additionally, the energies per link and per node are shown as well as the modularities of the respective partitions as calculated via the cavity method and from the q-state Potts model of Kanter and Sompolinsky (4.13) [22]. Also, numerical values obtained from maximizing Q_q using simulated annealing on networks of size $N = 10,000$.

q	2	3	4	5	6	7	8	9	10
η_1	0.333333	0.182277	0.120525	0.0880941	0.0684897	0.0555184	0.0463779	0.039630606	0.034469124
η_2	0.333333	0.106296	0.0487479	0.0270499	0.016716	0.0113865	0.00813282	0.006061892	0.004670888
η_3	0.0	0.134283	0.041858	0.0173799	0.0086019	0.00479192	0.00290601	0.001878418	0.001275993
η_4	0.0	0.0	0.0579794	0.017882	0.00697984	0.00319272	0.00163648	0.000913711	0.000545259
η_5	0.0	0.0	0.0	0.0258222	0.00793098	0.0029724	0.00128512	0.000618613	0.000323726
η_6	0.0	0.0	0.0	0.0	0.0116671	0.00530392	0.00130522	0.000541353	0.000248278
η_7	0.0	0.0	0.0	0.0	0.0	0.0	0.00162986	0.000582459	0.0002341
η_8	0.0	0.0	0.0	0.0	0.0	0.0	0.00241546	0.000743878	0.000262042
η_9	0.0	0.0	0.0	0.0	0.0	0.0	0.0	0.001099348	0.00033947
η_{10}	0.0	0.0	0.0	0.0	0.0	0.0	0.0	0.0	0.000499388
ΔE_1	−2.556	−2.369	−2.260	−2.187	−2.133	−2.092	−2.059	−2.031	−2.008
ΔE_2	−0.778	−0.684	−0.630	−0.593	−0.567	−0.546	−0.529	−0.516	−0.504
E_g	−1.389	−1.342	−1.315	−1.297	−1.283	−1.273	−1.265	−1.252	−1.252
Q_q	0.426	0.561	0.627	0.664	0.689	0.706	0.718	0.727	0.735
Q_q^{KS}	0.443	0.536	0.559	0.560	0.553	0.544	0.532	0.522	0.510
Experiment	0.417	0.550	0.610	0.646	0.666	0.680	0.691	0.697	0.704

6.1.6 Population Dynamics Approximation

After our treatment of two special cases, let us now turn to the general case of partitioning graphs of arbitrary degree distribution into arbitrarily many parts. The analytical results in the last subsection were obtained either for only two spin states or for a fixed excess degree of two. In more general cases we need to use a sampling technique known as population dynamics to find solutions for $P(\boldsymbol{h})$ and $\mathcal{Q}(\boldsymbol{u})$ [21].

Recall that the distribution of cavity biases $\mathcal{Q}(\boldsymbol{u})$ is entirely characterized by q order parameters η_τ. The $2^q - 1$ possible messages all have one of only q different probabilities of occurrence. Hence, instead of running a population dynamics algorithm on a population of $2^q - 1$ different messages, we can work with a population of q different order parameter indices only. Upon drawing an order parameter from the population, we then generate a message containing the appropriate number of non-zero entries at random. Every member τ of the population represents $q!/(q - \tau)!/\tau!$ possible messages. In practice, such an algorithm would run as follows:

1. Start with a population of order parameter indices $\tau \in \{1, ..., q\}$.
2. Draw an excess degree d from the distribution $q(d)$.
3. Draw d order parameter indices τ_i from the population at random and generate d messages $\boldsymbol{u}_i$ containing τ_i non-zero entries. All possible $q!/(q - \tau_i)!/\tau_i!$ messages have the same probability of being generated. This quenches the equipartition.
4. Calculate the cavity field $h_0 = \sum_{i=1}^{d} \boldsymbol{u}_i$ from these d messages and transform it into a message $\boldsymbol{u}_0 = \hat{u}(\boldsymbol{h_0})$.
5. Determine the order parameter index τ_0 of $\boldsymbol{u}_0$, i.e., count the number of ones in the message.
6. Replace an arbitrarily chosen order parameter index from the population by τ_0.
7. Continue with step 2 until convergence.

This converges to a population of order parameter indices in which every order parameter is found over-represented by a factor corresponding to its multiplicity. The symmetry condition is enforced in this algorithm by generating the appropriate messages for each order parameter randomly with equal probability.

Note that this procedure is especially suited for connected components of a graph. Since the distribution of order parameters is enforced to be symmetric, a balanced cut through all connected components is always enforced. This might not be the optimal partitioning, if the graph consists of more than one connected component [23, 24], but here, we are only interested in connected components, anyway.

6.2 Recovering Planted Cluster Structures

Until this point, we have been concerned with the maximum value of modularity that we may find in a purely random network. This value may serve as a reference value when reporting modularity scores on real world networks. But until now, we were not able to give an actual assessment of the (statistical) significance of the modularity that we have found in a real world network. We have seen that once we find a modularity score that exceeds the expectation value for an equivalent random graph, we can be rather confident that the modules we find are not the mere result of the clustering algorithms but represent actual reality in the data. But how confident can we really be? This section is devoted to answering this question. We will proceed by studying random graphs with an implanted modular structure and we will investigate to what extent this known cluster structure may be recovered by an algorithm maximizing modularity.

For our study, we will consider the following ensemble of random graphs with built-in cluster structure in the thermodynamic limit of infinitely large graphs: All nodes in the network belong to one of q pre-assigned classes or types. We consider equal-sized classes. The degree distribution $p(k)$ is the same across all types of nodes. The conditional probability that a link with a node of type $s \in (1, ..., q)$ on one end has a node of the same type on the other end, too, is denoted as $p(s|s) = p_{in}$, while that of having a node of a different type $r \neq s$ on the other end is denoted as $p(r|s) = p_{out} = (1-p_{in})/(q-1)$. This also implies that a fraction p_{in} of all edges in the network lies between nodes of the same type, while the remaining edges lie between nodes of different type. This ensemble of networks is entirely parametrized by the number of clusters q, the internal connection probability p_{in} and the degree distribution $p(k)$. We consider p_{in} in the range of $(1/q, 1)$ and thus can interpolate smoothly between completely random graphs without clusters for $p_{in} = 1/q$ and random graphs with built-in community structure with gradually denser clusters as p_{in} is increased up to the point of q disconnected components for $p_{in} = 1$ which form a trivially clustered graph.

Let us consider the built-in or "designed" modularity Q_q^d of such a network:

$$Q_q^d = p_{in} - \frac{1}{q}. \tag{6.47}$$

This relation follows from observing that the fraction of links within group s is simply $e_{ss} = p(s|s)/q$ independent of the degree distribution and that $a_s = \sum_r e_{rs} = 1/q \sum_r p(r|s)$ and by inserting these relations into the definition of Q. Now consider an assignment of Potts spin to the nodes in the network that corresponds exactly to the designed cluster structure or types, i.e., all nodes of the same type are in the same spins state and all pairs of nodes of different types are in different spin states. With such a configuration the Hamiltonian (6.1) would yield an energy per node of

$$E^p = -\frac{\langle k \rangle}{2} p_{in}, \tag{6.48}$$

which depends only on p_{in} independent of the shape of the degree distribution or the numbers of parts. Naturally, in the limit of $p_{in} \to 1$ we expect E^d to be the ground state energy and our pre-assigned cluster structure to represent a ground state configuration of (6.1), because the individual clusters are connected by only a few links. In this case, we can expect to recover the built-in cluster structure completely from the spin configuration in the ground state. In the limit of $p_{in} \to 1/q$ we expect the ground state configuration of (6.1) to be completely uncorrelated with the designed cluster structure, because the probability for a connection between nodes of the same type is the same as for nodes of different types and hence we cannot recover the built-in cluster structure. Between these two extremes, there exists a transition at which the designed cluster configuration starts to influence the ground state configuration. Since the goal of community detection or graph clustering is to recover a built-in, but unknown, cluster structure, this transition also marks the onset of detectable community structure. The naïve expectation is that we should be able to recover the designed structure when $E^d < E^{Rnd}$, i.e., when the cluster structure induces a minimum in the energy landscape that is lower than the ground state energy of the completely random network. We will show that this is only a rule of thumb and calculate the transition point exactly. We will further calculate the shape of the transition which will provide us with an expression for the accuracy of community detection.

We have seen that on our ensemble of random graphs with built-in cluster structure the problems of partitioning and clustering are identical. For convenience, we will hence study it in the language of graph partitioning, i.e., interactions between spins only along the edges of the graph plus the hard constraint of zero magnetization in the ground state.

We can make direct use of the cavity formulation developed earlier in this chapter. However, there is one more ingredient necessary before we are finally able to write the cavity equations for our clustered random graphs: We need to consider the different types of nodes. It is clear that for $p_{in} \approx 1$ the distribution of cavity fields and biases will depend on the type of node from which the message is sent. Let us define the distribution of incoming messages on nodes of type s as

$$\mathcal{Q}_{in}^s(\boldsymbol{u}) = \sum_r^q p(r|s)\mathcal{Q}^r(\boldsymbol{u}) = p_{in}\mathcal{Q}^s(\boldsymbol{u}) + \sum_{r \neq s} p_{out}\mathcal{Q}^r(\boldsymbol{u}). \tag{6.49}$$

This means that the distribution of the incoming messages on a node of type s is a mixture of the distributions of the outgoing messages of all types of nodes parametrized by the conditional probabilities of how the different types of nodes are connected in our model graph.

With this formulation at hand we are able to write the cavity equations for our model graph:

$$Q^s(\boldsymbol{u}) = \sum_{d=0}^{\infty} q(d) \int \prod_{i=1}^{d} (d^q \boldsymbol{u}_i Q_{in}^s(\boldsymbol{u}_i)) \, \delta \left(\boldsymbol{u} - \hat{u} \left(\sum_{i=1}^{d} \boldsymbol{u}_i \right) \right), \quad (6.50)$$

$$P_{\mathrm{cav}}^s(\boldsymbol{h}) = \sum_{d=0}^{\infty} q(d) \int \prod_{i=1}^{d} (d^q \boldsymbol{h}_i Q_{in}^s(\boldsymbol{u}_i)) \, \delta \left(\boldsymbol{h} - \sum_{i=1}^{d} \boldsymbol{u}_i \right), \quad (6.51)$$

$$P_{\mathrm{eff}}^s(\boldsymbol{h}) = \sum_{k=0}^{\infty} p(k) \int \prod_{i=1}^{k} (d^q \boldsymbol{u}_i Q_{in}^s(\boldsymbol{u}_i)) \, \delta \left(\boldsymbol{h} - \sum_{i=1}^{k} \boldsymbol{u}_i \right). \quad (6.52)$$

Compare these equations to those for the case of purely random networks (6.10–6.14). In principle, the different types of nodes may have a different degree distribution as well, but we set aside this further complication for now. The above formulation is in complete accordance with that presented in Ref. [25] for the vertex cover problem on networks with degree correlations where it was first used.

The expressions for the change in energy due to a site addition changes slightly to

$$\Delta E_1 = -\sum_s \frac{1}{q} \sum_{k=0}^{\infty} p(k) \int \prod_{i=1}^{k} (d^q \boldsymbol{u}_i Q_{in}^s(\boldsymbol{u}_i)) w \left(\sum_{i=1}^{k} \boldsymbol{u}_i \right) \quad (6.53)$$

$$= -\sum_s \frac{1}{q} \sum_{k=0}^{\infty} p(k) \int d^q \boldsymbol{h} P_{\mathrm{eff}}^s(\boldsymbol{h}) w(\boldsymbol{h}). \quad (6.54)$$

The factor $1/q$ is due to the fact that a fraction of $1/q$ of all nodes is of type s in our model of clustered random graphs. The energy needed for breaking a link is now written as

$$\Delta E_2 = \sum_{rs} e_{rs} \int d^q \boldsymbol{h}_1 d^q \boldsymbol{h}_2 P_{\mathrm{cav}}^r(\boldsymbol{h}_1) P_{\mathrm{cav}}^s(\boldsymbol{h}_2) \left(w(\boldsymbol{h}_1) - w(\boldsymbol{h}_1 + \hat{u}(\boldsymbol{h}_2)) \right)$$

$$= -\sum_{rs} \frac{p(r|s)}{q} \int d^q \boldsymbol{u}_1 d^q \boldsymbol{u}_2 Q^r(\boldsymbol{u}_1) Q^s(\boldsymbol{u}_2) \Theta(\boldsymbol{u}_1 \cdot \boldsymbol{u}_2), \quad (6.55)$$

where the factor $e_{rs} = p(r|s)/q$ is the fraction of edges running between nodes of type r and s, while Θ denotes the heaviside function with $\Theta(x) = 1$ for all $x > 0$ and zero otherwise. Again, ΔE_2 can be read as the probability of having an overlap in the messages $\boldsymbol{u}_1$ and $\boldsymbol{u}_2$. The ground state energy is calculated as before according to (6.17).

6.2.1 Symmetry Conditions

Our model of clustered random graphs has two limiting cases: $p_{in} \to 1/q$ and $p_{in} \to 1$ which correspond to a completely random network and a network made of q disconnected parts. First, we consider the case $p_{in} \to 1/q$. This corresponds to the case of a purely random network and has been treated in the previous section. We can then assume $\mathcal{Q}^s(\boldsymbol{u})$ does not depend on the type of node s from which the message is sent. From (6.49) we see immediately that there is no difference in the distribution of incoming and outgoing messages, i.e., $\mathcal{Q}_{in}(\boldsymbol{u}) = \mathcal{Q}(\boldsymbol{u})$ and we may choose an ansatz for $\mathcal{Q}$ as in (6.18).

Let us now consider the other limiting case, $p_{in} \to 1$. Without loss of generality, we can assume that the ground state is such that nodes of type s are in spin state s. Hence, the nodes of type s of the network send messages with only one non-zero entry in component s of $\boldsymbol{u}$, i.e., all q possible messages with only one non-zero entry are present in the network, but they are confined to the different connected parts. In other words, $\mathcal{Q}^s(\boldsymbol{u}) = \delta(\boldsymbol{u} - \boldsymbol{e}_s)$, with $\boldsymbol{e}_s$ the unit vector in direction s. Consider now the case $p_{in} < 1$, but still large enough, that the designed cluster structure is approximately the ground state. The probability that a node in part s sends a message with a single non-zero entry in the correct component is of course higher than sending a message with the non-zero entry in the wrong component $\mathcal{Q}^s(\boldsymbol{e}_s) > \mathcal{Q}^s(\boldsymbol{e}_{r \neq s})$. Additionally, $\mathcal{Q}^s(\boldsymbol{e}_{r \neq s})$ must be the same for all $r \neq s$ due to our equipartitioning constraint. The probability that a node of type s sends a message indicating a spin state r different from s must be the same for all $r \neq s$. In generalization of this argument, we see that $\mathcal{Q}^s(\boldsymbol{u})$ depends only on the number of non-zero entries in $\boldsymbol{u}$ and on whether $\boldsymbol{u}$ has a non-zero entry in the "correct" component s. The distribution of the remaining non-zero entries on the remaining $q-1$ "wrong" components must not matter for $\mathcal{Q}(\boldsymbol{u})$, as all of them are equivalent.

This leads us to suggest an ansatz involving $2q - 1$ order parameters η_{cw} with $c \in \{0,1\}$ denoting the number of correct non-zero entries and $w \in \{1-c, q-1\}$ denoting the number of wrong non-zero entries:

$$\mathcal{Q}^s(\boldsymbol{u}) = \eta_{cw} \text{ with } c = u^s \text{ and } w = \|\boldsymbol{u}\|^2 - 1. \tag{6.56}$$

Here, u^s denotes the sth component of the message vector $\boldsymbol{u}$ under consideration. The new order parameters obey the following normalization condition:

$$\sum_{c=0}^{1} \sum_{w=1-c}^{q-1} \binom{q-1}{w} \eta_{cw} = 1. \tag{6.57}$$

We have thus introduced a preferred spin orientation for the different types of nodes. Averaged over all types of nodes, however, we must ensure the absence of a preferred spin orientation. The probability of sending or receiving a message with τ non-zero entries from any type of nodes must only be a function of τ:

$$\sum_s \frac{1}{q} \mathcal{Q}^s(\boldsymbol{u}) = \sum_s \frac{1}{q} \mathcal{Q}_{in}^s(\boldsymbol{u}) = \eta_\tau \delta(\tau - \|\boldsymbol{u}\|^2). \tag{6.58}$$

Otherwise, the solution would not correspond to an equipartition. In other words, the condition for an equipartitioning is again enforced via demanding that a randomly drawn edge carries all messages with τ non-zero entries with equal probability. Note, however, that the distribution of outgoing and incoming messages may differ, $\mathfrak{Q}^s(\boldsymbol{u}) \neq \mathfrak{Q}_{in}^s(\boldsymbol{u})$, in general. The first equality in the above equation results from

$$\sum_s \frac{1}{q} \mathfrak{Q}_{in}^s(\boldsymbol{u}) = \sum_s \frac{1}{q} \sum_r p(r|s)\mathfrak{Q}^r(\boldsymbol{u}) = \sum_r \frac{1}{q}\mathfrak{Q}^r(\boldsymbol{u}) \sum_s p(r|s) = \sum_r \frac{1}{q}\mathfrak{Q}^r(\boldsymbol{u}).$$

(6.59)

We recover our earlier ansatz neglecting the different types of nodes by setting $\eta_{0,\tau} = \eta_{1,\tau-1} = \eta_\tau$ as a special case. We will see that this remains the only stable solution until p_{in} is larger than a critical value p_{in}^c above which $\eta_{0,\tau} < \eta_{1,\tau-1}$ is the stable solution, i.e., the probability that a node sends a message with a component indicating the "correct" direction is larger than the probability of sending a message without such a component.

With these conditions at hand, we can now turn to calculate η_{cw} as a function of p_{in}, the strength of the built-in clustering. The following development aims at determining the dependence of these order parameters on the properties of the network. We will proceed by examining graphs with arbitrary degree distributions but only two clusters and will then extend our analysis to larger numbers of clusters.

6.2.2 Graphs with Two Clusters

Let us first study networks with only two clusters or types of nodes A and B and therefore only two spin states. Before generalizing, we will carry out all calculations for a random Bethe lattice with only three links per node. Then we have three possible messages $\boldsymbol{u} \in \{(1,0), (0,1), (1,1)\}$. We use the abbreviations $\eta_{10}^s = \mathfrak{Q}^s(\boldsymbol{u} = (1,0))$, $\eta_{01}^s = \mathfrak{Q}^s(\boldsymbol{u} = (0,1))$ and $\eta_{11}^s = \mathfrak{Q}^s(\boldsymbol{u} = (1,1))$. The cavity equation (6.50) for the distribution of messages $\mathfrak{Q}^s(\boldsymbol{u})$ can then be written for each of the two types of nodes $s \in \{A, B\}$ as the following system of three equations:

$$\eta_{11}^{A/B} = \underbrace{(p_{in}\eta_{11}^{A/B} + p_{out}\eta_{11}^{B/A})^2}_{\eta_{11}^{A/B,in}}$$

(6.60)

$$+2\underbrace{(p_{in}\eta_{10}^{A/B} + p_{out}\eta_{10}^{B/A})}_{\eta_{10}^{A/B,in}}\underbrace{(p_{in}\eta_{01}^{A/B} + p_{out}\eta_{01}^{B/A})}_{\eta_{01}^{A/B,in}},$$

$$\eta_{10}^{A/B} = (\eta_{10}^{A/B,in})^2 + 2\eta_{10}^{A/B,in}\eta_{11}^{A/B,in},$$

(6.61)

$$\eta_{01}^{A/B} = (\eta_{01}^{A/B,in})^2 + 2\eta_{01}^{A/B,in}\eta_{11}^{A/B,in}.$$

(6.62)

The symmetry condition (6.58) then reads

$$\eta_{11}^A + \eta_{11}^B = \eta_{11}^{A,in} + \eta_{11}^{B,in} = 2\eta_2,$$

(6.63)

$$\eta_{10}^A + \eta_{10}^B = \eta_{10}^{A,in} + \eta_{10}^{B,in} = 2\eta_1,$$

(6.64)

$$\eta_{01}^A + \eta_{01}^B = \eta_{01}^{A,in} + \eta_{01}^{B,in} = 2\eta_1.$$

(6.65)

We see that we have in principle three messages for each of the two types of nodes which means we have six order parameters to determine. Let us compare the Eqs. (6.61) and (6.62) for both types of nodes A and B. They are invariant under an interchange of A with B if at the same time the subscripts 10 and 01 are interchanged. This suggests a solution in which $\eta_{10}^A = \eta_{01}^B = \eta_{10}$ and $\eta_{01}^A = \eta_{10}^B = \eta_{01}$ and consequently $\eta_{11}^A = \eta_{11}^B = \eta_{11}$. Thus, we are left with only two free order parameters to determine. With the introduction of the new order parameters η_{10}, η_{01}, η_{11} we have, without loss of generality, assigned a preferred spin orientation for the nodes of type A and B. The new order parameters have precisely the form η_{cw} introduced on general grounds in our discussion about the symmetry of the solution to the field equations. We hence have assigned direction $(1, 0)$ as "correct" for node type A and direction $(0, 1)$ as "correct" for node type B. The six Eqs. (6.61) and (6.62) can now be written as only three equations, simply by dropping the indices A and B.

For our random Bethe lattice with three links per node Fig. 6.4 shows the order parameters as solutions of (6.61) and (6.62), the ground state energy of (6.1) according to (6.17) and the accuracy of recovering the built-in cluster structure as a function of the intra-cluster link probability p_{in}. Here we have defined the accuracy of recovering the built-in cluster structure as a function of the effective fields calculated according to (6.52). We denote as accuracy the fraction of nodes from a pre-assigned cluster s that see a maximum component of the effective field in direction s. If that maximum is n-fold degenerate, we assume the node is assigned into any of the n corresponding clusters with equal probability. We can read it directly as percentage of nodes classified correctly.

In our example we find a ground state energy of $E^{Rnd} = -25/18 \approx -1.39$ for the completely random lattice at $p_{in} = 1/2$ from the order parameters $\eta_{11} = \eta_{10} = \eta_{01} = 1/3$. This means that even in this completely random case, we find a ground state configuration with an empirical internal link probability of $p_{in}^{emp} = 25/27 \approx 0.93$. Naïvely, we expect changes in the ground state energy to happen when $p_{in} > p_{in}^{emp}$ or equivalently, when $E^{Rnd} > E^d$. From Fig. 6.4 we now see that the ground state energy starts to change already at $p_{in} = p_{in}^c < p_{in}^{emp}$. We see that the true ground state energy is always smaller than or equal to the planted ground state energy $E \leq E^p$ where the equal sign holds only for $p_{in} = 1$. This means we are always able to optimize the cut around the designed partition and that the designed partition is never optimal except for $p_{in} = 1$. However, we also see that beyond p_{in}^c the optimization makes only small changes to the built-in assignment of nodes into clusters. The point at which the accuracy increases is indicated by $E < E^{Rnd}$, i.e., we find a lower ground state energy than we expect to find in a completely random network.

The most striking and possibly unexpected feature is that the accuracy does not increase smoothly between $1/q < p_{in} < 1$. There exist cluster structures which may be very pronounced but which we cannot recover with our approach. These remain hidden behind alternative assignments of spin states

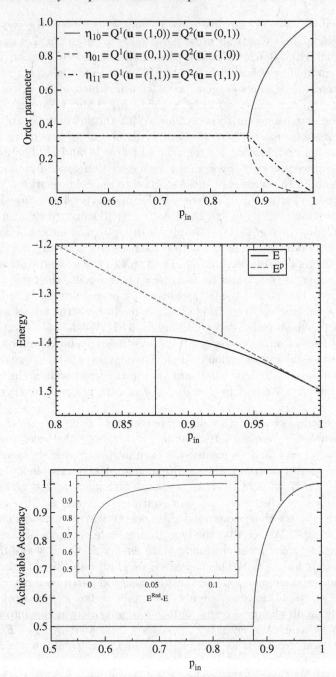

Fig. 6.4. *Top:* The order parameters η_{cw} for the planted bisection problem on a random Bethe lattice with $k = 3$ links per node as a function of p_{in}. The planted cluster structure in the network does not influence the ground state configuration

to nodes which produce lower energies and are completely uncorrelated with the built-in cluster structure. The existence of these "spurious" solutions to the clustering problem in networks shows that different algorithms must also fail at recovering the pre-assigned clusters. Nevertheless, all partitions found "feign" an internal link probability p_{in} of at least p_{in}^{emp}.

Let us now generalize these results to graphs with two clusters and arbitrary degree distribution. Note that so far we have only introduced a notation that takes the symmetry conditions we have derived into account and we can use this notation to generalize the cavity equations (6.61) and (6.62) to any desired degree distributions:

$$\eta_{11} = \sum_{n0=0}^{\infty} \sum_{n=0}^{\infty} q(n_0 + 2n) \frac{(n_0 + 2n)!}{n_0! n! n!} \left(\eta_{10}^{in}\right)^n \left(\eta_{01}^{in}\right)^n \left(\eta_{11}^{in}\right)^{n_0}, \tag{6.66}$$

$$\eta_{10} = \sum_{n0=0}^{\infty} \sum_{n_1 > n_2}^{\infty} q(n_0 + n_1 + n_2) \frac{(n_0 + n_1 + n_2)!}{n_0! n_1! n_2!} \left(\eta_{10}^{in}\right)^{n_1} \left(\eta_{01}^{in}\right)^{n_2} \left(\eta_{11}^{in}\right)^{n_0}, \tag{6.67}$$

$$\eta_{01} = \sum_{n0=0}^{\infty} \sum_{n_1 > n_2}^{\infty} q(n_0 + n_1 + n_2) \frac{(n_0 + n_1 + n_2)!}{n_0! n_1! n_2!} \left(\eta_{01}^{in}\right)^{n_1} \left(\eta_{10}^{in}\right)^{n_2} \left(\eta_{11}^{in}\right)^{n_0}. \tag{6.68}$$

We can easily solve these equations for every value of p_{in} and every degree distribution by iteration. Recall that for values of $p_{in} < p_{in}^c$ the solution is that of the unclustered case, i.e., $\eta_{10} = \eta_{01} = \eta_1$. This means that in this parameter range, nodes of type A are equally likely to send a message pointing in direction $(1, 0)$ as in direction $(0, 1)$. The cluster structure of the graph does not play a role for the distribution of messages. Our goal is now to calculate the critical value of p_{in}^c beyond which the cluster structure of the graph under study starts to influence the ground state structure. To do so, we will need the solution of the unclustered case, i.e., for $p_{in} = 1/q$ which corresponds to the original graph partitioning problem which we have solved in the previous section.

6.2.3 Onset of Detectable Cluster Structure

We now proceed to the calculation of the critical value of p_{in}^c beyond which the designed cluster structure starts to influence the ground state of the partitioning problem. To do so, we first rewrite the order parameters slightly by setting $\eta_{10} = \eta_{01} + \delta$. From the definition of $Q_{in}(\boldsymbol{u})$ in (6.49) we can write

---

Fig. 6.4. (*Continued*) until a critical value of p_{in} is reached. *Middle:* The ground state energy E of (6.1) and the energy of the planted cluster structure E^p vs. p_{in}. The left vertical blue line indicates the critical value of p_{in}^c beyond which $\eta_{10} > \eta_{01}$ and $E < E^{Rnd}$ and the planted cluster structure starts to influence the ground state energy. The right vertical blue line indicates the naïve value of $p_{in}^n = 2E^{Rnd}/\langle k \rangle$ beyond which $E^p < E^{Rnd}$. *Bottom:* The accuracy with which the planted cluster structure may be recovered. Again, the two vertical lines indicate p_{in}^c and p_{in}^n. The inset shows that the accuracy increases dramatically, as soon as $E < E^{Rnd}$.

$$\eta_{10}^{in} = \eta_{10} - \delta p_{out}, \tag{6.69}$$

$$\eta_{01}^{in} = \eta_{10} - \delta p_{in}. \tag{6.70}$$

Inserting these relations into (6.67) and (6.68) and linearizing we find an equation for δ:

$$\delta = \sum_{n_0=0}^{\infty} \sum_{n_1 > n_2}^{\infty} q(n_0+n_1+n_2)\frac{(n_0 + n_1 + n_2)!}{n_0!n_1!n_2!}(n_1-n_2)(p_{in}-p_{out})\delta\eta_{10}^{n_1+n_2-1}\eta_{11}^{n_0}. \tag{6.71}$$

We are interested in the largest value of p_{in} at which δ is non-zero but we can still approximate the order parameters as $\eta_{10} = \eta_{01} = \eta_1$ and $\eta_{11} = \eta_2$, i.e., with the solutions from the completely random graph. This is the case for

$$(p_{in}^c - p_{out}^c)^{-1} = \sum_{n_0=0} \sum_{n_1 > n_2} q(n_0 + n_1 + n_2)\frac{(n_0 + n_1 + n_2)!}{n_0!n_1!n_2!}(n_1 - n_2)$$
$$\times \eta_1^{n_1+n_2-1}\eta_2^{n_0}. \tag{6.72}$$

Note how the order parameters for the unclustered case enter into the calculation. For a Bethe lattice with $q(d) = \delta(d-2)$, i.e., three connections per node, and solutions to the unclustered problem of $\eta_1 = \eta_2 = 1/3$, we find $p_{in}^c = 7/8$ as we could also have read from Fig. 6.4. Note the similarity of (6.72) with the definitions of the order parameter η_2 and X in (6.33). For graphs with a Poissonian degree distribution with mean λ, we can simplify (6.72) further, because of the equivalence of degree distribution $p(k)$ and excess degree distribution $q(d)$:

$$(p_{in}^c - p_{out}^c)^{-1} = \lambda \left(\eta_2 + \frac{X_\lambda}{\eta_1}\right). \tag{6.73}$$

In the above equations, it is understood that $p_{in} + p_{out} = 1$.

Let us now compare the critical value p_{in}^c for networks with two clusters across different network densities and different degree distributions. As before, we will compare graphs with Poissonian degree distribution (ER) and the two types of scale free networks (SF Δk and SF k_{min}) already introduced in Sect. 5.2. Figure 6.5 shows a comparison of p_{in}^c for these networks for different realizations with different average degree. It is interesting to note the correspondence with Fig. 6.2. The critical point p_{in}^c for a network with given average connectivity is lowest for the degree distribution which is hardest to cut, i.e., shows the lowest modularity in an unclustered case. The reason for this is intuitive. The ground state energy of the partitioning Hamiltonian in the completely unclustered case is just the depth of the deepest valley in this energy landscape. If these are shallow, i.e., the graph is difficult to cut and exhibits low values of modularity, then the minima induced by the designed cluster structure will be more easily recognized. The general pattern across all degree distributions is that p_{in} decreases with increasing average connectivity and the differences between the different degree distributions matter less. Clusters are easier to detect in denser graphs.

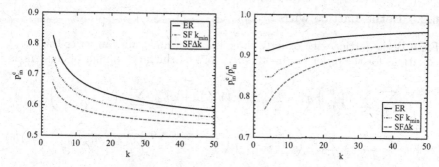

Fig. 6.5. *Left:* The critical value of p_{in} beyond which the cluster structure starts to influence the ground state of the bisection problem, i.e., below which clusters cannot be detected. We compare Erdös–Renyi graphs (ER) with a Poissonian degree distribution $p(k) = e^{-\lambda}\lambda^k/k!$ and two types of scale-free degree distributions. The first one being a stretched power law (SF Δk) of the form $p(k) = (k + \Delta k)^{-\gamma}$ with $\Delta k \in [18, 19]$ and the second (SF $k_{\min}$) being of the form $p(k) = k^{-\gamma}$ with a varying minimum degree $k_{\min}$ with $k_{\min} \in [19, 20]$. For both scale-free distributions we choose $\gamma = 3$. Since we are interested only in the behavior of the giant connected component, we set $p(k = 0) = 0$ in all cases. Note the correspondence to Fig. 6.2. The critical value of p_{in} is smaller, i.e., clusters are easier to detect, for networks with degree distributions which are harder to cut, i.e., which display lower modularities in the unclustered case. *Right:* The ratio of p_{in}^c and p_{in}^n, the naïve estimate for the transition point $p_{in}^n = 2E^{Rnd}/\langle k \rangle$ which always overestimates p_{in}^c.

6.2.4 Graphs with More than Two Clusters

After we have dealt with graphs containing only two clusters, we will now turn to the problem of networks with more than two clusters. While in the case of two clusters we could write the cavity equations for an arbitrary degree distribution, we were able to find a simple formulation for more than two clusters only for one special topology, the random Bethe lattice with three links per node, such that a cavity field is composed of combining only two messages. For graphs with a different topology, we will have to resort to a population dynamics algorithm to be described in the next section.

Let us first consider how to calculate the η_{cw}^{in} for general q. A message with $\tau = c + w$ non-zero entries will have one correct entry and w wrong entries for τ types of nodes, while for $q - \tau$ types of nodes, it will have no correct entry but $w + 1$ wrong entries. Hence we can write straightforwardly

$$\eta_{1w}^{in} = p_{in}\eta_{1w} + \frac{(1 - p_{in})}{q - 1}\left(w\eta_{1w} + (q - 1 - w)\eta_{0,w+1}\right), \qquad (6.74)$$

$$\eta_{0w}^{in} = p_{in}\eta_{0w} + \frac{(1 - p_{in})}{q - 1}\left(w\eta_{1,w-1} + (q - 1 - w)\eta_{0,w}\right). \qquad (6.75)$$

We see that this recovers the unclustered case if $\eta_{1w} = \eta_{0,w+1} = \eta_\tau$.

6.2.5 Bethe Lattice with k = 3

For a Bethe lattice with exactly three links per node, we can write the cavity equations for the order parameters η_{cw} for an arbitrary number of clusters as

$$
\eta_{1w} = \sum_{c=0}^{1} \sum_{\alpha=1-c}^{w-c} \binom{w}{\alpha} \eta_{c,\alpha}^{in} \eta_{1-c,w-\alpha}^{in} + \left(\eta_{1w}^{in} \right)^2 + 2\eta_{1w}^{in} \sum_{\alpha=1}^{q-1-w} \binom{q-1-w}{\alpha} \eta_{1,w+\alpha}^{in}
$$

$$
+ \sum_{\alpha=1}^{q-1-w} \sum_{\beta=1}^{q-1-w-\alpha} \binom{q-1-w}{\alpha} \binom{q-1-w-\alpha}{\beta} \eta_{1,w+\alpha}^{in} \eta_{1,w+\beta}^{in}, \tag{6.76}
$$

$$
\eta_{0w} = \sum_{\alpha=1}^{w-1} \binom{w}{\alpha} \eta_{0,\alpha}^{in} \eta_{0,w-\alpha}^{in} + \left(\eta_{0w}^{in} \right)^2 + 2\eta_{1w}^{in} \eta_{0w}^{in}
$$

$$
+2 \sum_{\alpha=1}^{q-1-w} \binom{q-1-w}{\alpha} (\eta_{0w}^{in} \eta_{1,w+\alpha}^{in} + \eta_{1w}^{in} \eta_{0,w+\alpha}^{in} + \eta_{0w}^{in} \eta_{0,w+\alpha}^{in})
$$

$$
+ \sum_{\alpha=1}^{q-1-w} \sum_{\beta=1}^{q-1-w-\alpha} \binom{q-1-w}{\alpha} \binom{q-1-w-\alpha}{\beta}
$$

$$
\times (\eta_{0,w+\alpha}^{in} \eta_{1,w+\beta}^{in} + \eta_{1,w+\alpha}^{in} \eta_{0,w+\beta}^{in} + \eta_{0,w+\alpha}^{in} \eta_{0,w+\beta}^{in}). \tag{6.77}
$$

These equations must be read as describing the different ways and their probabilities in which two messages can be combined to form a cavity field and the resulting cavity bias. They can be iterated very easily to give the order parameters over the entire range of p_{in}.

Figure 6.6 shows two examples for a three and a four partitioning of the Bethe lattice with three links per node. Note the similarity also to Fig. 6.4 and that p_{in}^c decreases slightly for increasing q. The qualitative behavior of the order parameters is the same across all degree distributions. Note how $\eta_{1,w-1} = \eta_{0,w}$ for all $p_{in} \leq p_{in}^c$ and $\eta_{1,w-1} > \eta_{0,w}$ for all $p_{in} > p_{in}^c$. Also note the different scales of the graphs, η_{10} quickly starts to dominate all other order parameters.

6.2.6 Population Dynamics Formulation of the Cavity Equations

To generalize our calculation to other network topologies, we have to use a population dynamics algorithm in order to solve the field equations. Recall that the distribution of cavity biases $\mathfrak{Q}(\boldsymbol{u})$ is entirely characterized by $2q - 1$ order parameters η_{cw}. The $2^q - 1$ possible messages all have one of only $2q - 1$ different probabilities of occurrence. Hence, instead of running a population dynamics algorithm on a population of $2^q - 1$ different messages, we can work with a population of $2q - 1$ different pairs of order parameter indices (c, w) only. Upon drawing an order parameter index pair from the population, we then generate a message containing the appropriate number of correct/wrong zero and non-zero entries at random. In practice, such an algorithm would run as follows:

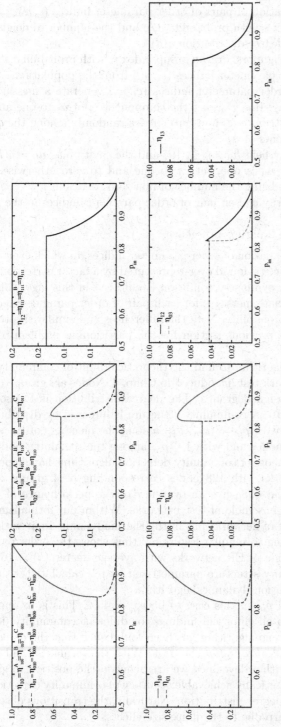

Fig. 6.6. The $2q-1$ order parameters for the q-partitioning problem in a random Bethe lattice with three links per node. *Top:* From left to right the five different order parameters for $q=3$ clusters. As before, we have used the abbreviation $\eta_{100}^A = Q^A(\boldsymbol{u}=(1,0,0))$ etc., for the three clusters present by design in the network. *Bottom:* The seven different order parameters for $q=4$ clusters. Note how order parameters can be grouped into pairs with $c+w$ below which the network-inherent cluster structure does not influence the ground state configuration. This p_{in} is smaller for higher numbers of q. Also compare the case of the bi-partition in Fig. 6.4.

1. Start from a population of pairs of order parameter indices (c, w).
2. Draw a group index s with probability $1/q$ and the number of neighbors d from the excess degree distribution $q(d)$.
3. For each of the d neighbors, draw a group index r_i with probability $p(r_i|s)$ and a pair of order parameter indices (c_i, w_i) from the population.
4. For each pair of order parameter indices (c_i, w_i) generate a message u_i in the following way: If $c_i = 1$, set the component r_i of u_i to one and to zero otherwise. Distribute w_i non-zero entries randomly among the $q - 1$ remaining components of u_i.
5. Calculate the cavity field $h_0 = \sum_{i=1}^{d} u_i$ and the cavity bias $u_0 = \hat{u}(h_0)$.
6. If component s of u_0 is one, set c_0 to one and to zero otherwise. Set $w_0 = \tau - c_0$ with τ being the number of non-zero entries in u_0.
7. Replace an arbitrarily chosen pair of order parameter indices in the population by (c_0, w_0).
8. Continue with step 2 until convergence.

This converges to a population of order parameter indices, in which every pair of order parameter indices is found over-represented by a factor corresponding to its multiplicity. Our symmetry condition is enforced in this algorithm by generating the appropriate messages for each pair of order parameter indices randomly with equal probability. Note the differences and similarities to the algorithm given in the previous section for the partitioning problem of the purely random network.

Let us apply this algorithm to find the theoretical limit of community detection for the benchmark test introduced in Chap. 4. Nodes are grouped into four equal-sized pre-assigned groups. The degree distribution is Poissonian with a mean of $\langle k \rangle = 16$. A community structure is imposed by distributing the links of each node with $p_{in} = \langle k_{in} \rangle / \langle k \rangle$ among the members of the same pre-assigned group of nodes and with $1 - p_{in}$ among the remaining nodes in the networks. Most authors of community detection algorithms have reported their results for networks with 128 nodes corresponding to 4 groups of 32 nodes each. Danon et al. [28] give an overview of the performance of various algorithms on this ensemble of test networks. Within our formalism, we consider the same class of networks but in the thermodynamic limit with the number of nodes tending to infinity. Figure 6.7 thus shows the order parameter for a $q = 4$ partition of ER networks with average degree $\langle k \rangle = 16$ and a predesigned community structure parametrized by p_{in} calculated with the above-described population dynamics algorithm.

We find the critical p_{in} in this case to be $p_{in}^c \approx 45\%$. This is in remarkable correspondence to all empirical findings for different community detection algorithms: So far, no algorithm has been found such that the detection accuracy does not break down at $p_{in} \approx 45\%$. With the formalism and population dynamics algorithm developed we are hence in the position to give a limit curve for the theoretically achievable accuracy of community detection in graphs with arbitrary degree distribution. Naturally, the transitions observed in Ref. [28] are not sharp due to the finite size effects.

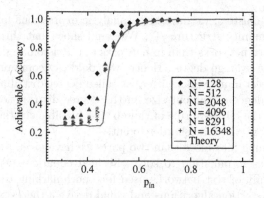

Fig. 6.7. Achievable accuracy for the planted partition problem on ER graphs with $N \to \infty$, $\langle k \rangle = 16$ and four equal-sized clusters and numerical results obtained from the best graph clustering algorithms on equivalent networks with $N = 128$ to $N = 16348$ nodes [26, 28]. The observed differences are attributed to finite size effects. From the figure we read off a critical $p_{in}^c \approx 45\%$. With this value we recover the critical value obtained by the best community detection algorithms [28] and at the same time have found the theoretical limit for community detection algorithms on this benchmark.

6.3 Conclusion

In this chapter we have used the cavity method directly at zero temperature to study the problems of partitioning and clustering in random graphs with arbitrary degree distributions. In particular, we studied the problem on random graphs with a built-in cluster or community structure and investigated to what extent such built-in structure influences the minimum cost solution of the partitioning problem. Only if the built-in cluster structure influences the partitioning problem, we can hope to recover a possibly hidden cluster structure in real world networks. The restriction to the case of equal-sized clusters leads to a symmetry condition on the solution of the problem. It allowed a reduction of $2^q - 1$ order parameters to only $2q - 1$. We could show that the number of order parameters is further reduced to q if the built-in cluster structure does not influence the minimum cost solution of the partitioning problem.

For the case of the bi-partitioning problem, we could derive analytical formulas for the quality of the optimal partitioning solution. In the limit of dense random graphs we could recover results known from replica calculations and show that cut size is smaller in networks with Poissonian degree distribution than in networks with scale-free degree distributions and the same average degree. We showed that network-inherent cluster structure can only lead to solutions of the field equations different from the case of random graphs without cluster structure, if p_{in} exceeds a critical value p_{in}^c which thus marks the threshold of detectable cluster or community structure. For the case of networks with two clusters, we have given analytic expressions for this onset of detectable community structure p_{in}^c. We could show that this critical p_{in}^c is lower in scale-free networks than in networks of a Poissonian degree distribution of the same average degree. Hence, we could say community structures are easier to detect in networks with a fat-tailed degree distribution. Together with the dependence of the cut size on the degree distribution, this can be summarized by the intuitive rule of thumb that possible clusters are easier to detect in networks which are harder to cut.

For partitioning into more than two parts we have given an efficient population dynamics formulation to solve the field equations. We applied our results to a family of test networks used for benchmarking the performance of community detection algorithms and could derive a theoretical limit curve for the possibly achievable detection accuracy of any community detection algorithm. The comparison with the performance of algorithms published in the literature [28] has shown that this limit has already been reached by the best available algorithms.

The key result of our analysis is the demonstration that possibly strong clusters exist in a network which are completely hidden behind alternative spurious solutions. This is a typical limitation for exploratory data analysis or data-driven research into which no prior knowledge enters. In order to increase the detection accuracy of graph clustering algorithms, developers should hence search for ways of incorporating prior knowledge into community detection algorithms.

Our work has a number of implications for the analysis of real world networks with community detection algorithms. It could be shown that the detection accuracy increases very fast as soon as the ground state energy (resp. modularity) exceeds the expectation values for completely random graphs. Hence, practitioners looking for means of assessing the possible accuracy of their analysis may use our results as a baseline. More research is needed on the sample to sample variation due to finite size effects in networks before quantitative assertions can be made.

On the other hand, we have shown that modularities which do not exceed the expectation values do not imply that no cluster structure can exist, but rather that the cluster structures found can be considered uncorrelated with any true cluster structure. Even though cluster structure may be present, it may remain undetectable and hidden behind alternative solutions to the clustering problem that have zero correlation with the true solution. If we were to draw an analogy to unsupervised learning problems on multivariate data, we could say the average connectivity of a network plays the role of the ratio α between the number of data points and the dimensionality of a multivariate data set. A number of transitions from unrecoverable to recoverable cluster structure have been observed for multivariate data [27, 29–32] but always in the number of data points. This means, for a data set of dimensionality D, there is a minimum number of data points αD necessary to be able to recover a built-in cluster structure. Hence, it is possible to recover any cluster structure, as long as enough data are provided. The fundamental difference in the case of clustering in networks is that the average connectivity is not a free parameter in sparse networks and cannot be increased by adding more nodes to the network. Adding nodes to the network inevitably increases the dimensionality of the data. Thus we are dealing with a qualitatively different phenomenon. Though we have only derived these results here for the case of community structures, i.e., diagonal block models, with equal-sized blocks, they remain, at least in principle, valid also for other types of block models. They may be valuable for the design of network clustering algorithms and their benchmarks as well as for a critical assessment of the amount of information that can be derived from unsupervised learning or data mining on networks.

References

1. F. Y. Wu. The Potts model. *Reviews of Modern Physics*, 54(1):235–368, 1982.
2. L. Viana and A. J. Bray. Phase diagrams for dilute spin glasses. *Journal of Physics C: Solid State Physics*, 18:3037–3051, 1985.
3. K. Y. M. Wong and D. Sherrington. Graph bipartitioning and spin glasses on a random network of finite valence. *Journal of Physics A: Mathematical and General*, 20:L793–L799, 1987.
4. M. Mezard and G. Parisi. Mean-field theory of randomly frustrated systems with finite connectivity. *Europhysics Letters*, 3(10):1067–1074, 1987.
5. I. Kanter and H. Sompolinsky. Mean-field theory of spin-glasses with finite co-ordination number. *Physical Review Letters*, 58(2):164–167, 1987.
6. D. Sherrington and K. Y. M. Wong. Graph bipartitioning and the bethe spin glass. *Journal of Physics A: Mathematical and General*, 20:L785–L791, 1987.
7. Y. Y. Goldschmidt and C. De Dominicis. Replica symmetry breaking in the spin-glass model on lattices with finite connectivity: Application to graph partitioning. *Physical Review B*, 410(4):2184–2197, 1990.
8. M. Mezard and G. Parisi. The cavity method at zero temperature. *Journal of Statistical Physics*, 111(1/2):1–34, 2003.
9. A. Braunstein, R. Mulet, A. Pagnani, M. Weigt, and R. Zecchina. Polynomial iterative algorithms for coloring and analyzing random graphs. *Physical Review E*, 68:036702, 2003.
10. R. Mulet, A. Pagnani, M. Weigt, and R. Zecchina. Coloring random graphs. *Physical Review Letters*, 89, 2002.
11. M. Mezard and G. Parisi. The bethe lattice spin glass revisited. *European Physical Journal B*, 20:217–233, 2001.
12. M. E. J. Newman. Mixing patterns in networks. *Physical Review E*, 67:026126, 2003.
13. J. R. Banavar, D. Sherrington, and N. Sourlas. Graph bipartioning and statistical mechanics. *Journal of Physics A: Mathematical and General*, 20:L1–L8, 1987.
14. M. J. de Oliveira. Graph optimization problems on the bethe lattice. *Journal of Statistical Physics*, 54(1/2):477–493, 1989.
15. Y. Fu and P. W. Anderson. Application of statistical mechanics to NP-complete problems in combinatorial optimisation. *Journal of Physics A: Mathematical and General*, 19:1605–1620, 1986.

16. P. Erdős and A. Rényi. On the evolution of random graphs. *Publications of the Mathematical Institute of the Hungarian Academy of Sciences*, 5:17–61, 1960.
17. E. T. Jaynes. Information theory and statistical mechanics ii. *Physical Review*, 108(2):171–190, 1957.
18. G. Palla, I. Derenyi, I. Farkas, and T. Vicsek. Uncovering the overlapping community structure of complex networks in nature and society. *Nature*, 435:814, 2005.
19. E. T. Jaynes. Information theory and statistical mechanics. *Physical Review*, 106(4):620–630, 1957.
20. E. A. Bender and E. R. Canfield. The asymptotic number of labeled graphs with given degree distribution. *Journal of Combinatorial Theory A*, 24:296, 1978.
21. A. K. Hartmann and H. Rieger, editors. *New Optimization Algorithms in Physics*. Wiley-Vch, Weinheim, 2004.
22. I. Kanter and H. Sompolinsky. Graph optimisation problems and the potts glass. *Journal of Physics A: Mathematical and General*, 20:L636–679, 1987.
23. W. Liao. The graph-bipartitioning problem. *Physical Review Letters*, 59(15):1625–1628, 1987.
24. W. Liao. Replica-symmetric solution of the graph-bipartitioing problem. *Physical Review A*, 37:587–595, 1988.
25. A. Vázquez and M. Weigt. Computational complexity arising from degree correlations in networks. *Physical Review E*, 67:027101, 2003.
26. R. Guimera and L. A. N. Amaral. Functional cartography of complex metabolic networks. *Nature*, 433:895–900, 2005.
27. A. Engel and C. Van den Broeck. *Statistical Mechanics of Learning*. Cambridge University Press, New York, 2001.
28. L. Danon, J. Dutch, A. Arenas, and A. Diaz-Guilera. Comparing community structure indentification. *Journal of Statistical Mechanics*, P09008, 2005.
29. M. Biehl and A. Mietzner. Statistical mechanics of unsupervised learning. *Europhysics Letters*, 24(5):421–426, 1993.
30. C. Van den Broeck and P. Reimann. Unsupervised learning by examples: Online versus off-line. *Physical Review Letters*, 76(12):2188–2191, 1996.
31. P. Reimann and C. Van den Broeck. Learning by examples from a nonuniform distribution. *Physical Review E*, 53(4):3989–3998, 1996.
32. A. Buhot and M. B. Gordon. Phase transitions in optimal unsupervised learning. *Physical Review E*, 57(3):3326–3333, 1998.

7

Applications

The previous chapters have mainly focussed on the theoretical aspects of our block modeling procedure. This chapter now is devoted to applications. The first will be an analysis of the United Nations commodity trade database on the level of individual countries. We aim at inferring the block model which best captures the flow of trade between different groups of countries. With this example we have a rather dense weighted network. The block model we infer will hence show where the volume of trade exceeds the expectation values based on the total import and export volumes of individual countries, in other words, "preferred" trade relations. The second example will show the analysis of a very sparse data set of consumer interactions on an online auction site. We are looking for groups of customers with common interests and will hence try to find the best fit of a diagonal block model to the network.

7.1 Block Modeling the World Trade Network

In order to demonstrate our block modeling approach, we investigate a data set for the year 2000 from the United Nations commodity trade database [1]. Independent research [2,3] has shown that the 55 commodities that make up the bulk of world trade, when factor analyzed, form 5 major groups and that commodities are highly correlated within each group. These groups are differentiated by proportions of production with extraction, capital-intensive or labor-intensive processing. The five groups of commodities are (a) food products and by-products, (b) simple extractive, (c) sophisticated extractive, (d) high technology and heavy manufacture and (e) low wage/light manufacture. Representative for each of these groups, we chose one commodity each and obtained five different networks of commodity trade. The five commodities are (a) meat and meat preparations, (b) animal oil and fats, (c) paper, paperboard and articles of pulp, (d) machinery and (e) footwear. The data set is based on the volumes of import as reported by 112 countries to the UN in 2000. The only pretreatment applied to the data was to take the logarithm of

Reichardt, J.: *Applications*. Lect. Notes Phys. **766**, 119–147 (2009)
DOI 10.1007/978-3-540-87833-9_7   © Springer-Verlag Berlin Heidelberg 2009

the trade volumes which preserves the relative strength of trade volumes but reduces the effect that the fit of high volume countries alone dominates the quality of the role models.

Since the five different commodities had been found to be largely independent [2] and also have different overall volumes, we do not simply sum the volumes but extend (3.15) in order to accommodate for different types of links in the network. Instead of performing the same analysis for the different commodities independently and trying to form a consensus a posteriori, we include the different kinds of traded goods at the same time in the model finding process. The quantity that we maximize is

$$Q(\{\sigma\}) = \frac{1}{2} \sum_c \sum_{r,s}^q \|m_{rs}^c - [m_{rs}^c]\|. \tag{7.1}$$

Here, the first sum runs over the different commodities c and every country i is assigned exactly one role $\sigma_i \in \{1, ..., q\}$ which it assumes in all block models. Further, m_{rs}^c is the log of the total volume of commodity c imported by countries in role r from those in role s. For the calculation of $[m_{rs}^c]$ we use (3.22), i.e., the expectation value for the trade between two countries is based on the marginals. It is hence proportional to the product of the log of the total import volume of countries in role r and the log of the total export volume of countries is role s. Once an assignment of roles to countries has been found that maximizes (7.1), we can read off the five different image graphs B_{rs}^c directly from the terms $m_{rs}^c - [m_{rs}^c]$ as before. The different models can then be overlaid easily as the same countries are assigned into the same roles for all of them. The computational effort for this multi-commodity block modeling is still moderate as it increases over the case of one link type only by a factor of the number of different commodities.

Before discussing the block models we obtain, we need to determine the optimal number of roles. We calculate Q_{max}^c for each of the five commodities separately according to (3.18). For different numbers of roles q, we then maximize (7.1) and find $Q(q)/Q_{max}$ averaged over the five commodities. This is necessary since we can define Q_{max} only for a single link type and (7.1) aims at constructing a consensus model for all link types. This average value tells us what fraction of the total link structure we mimic in our image graph. As a random null model, we created randomized versions of the empirical data by rewiring the original network but keeping the number of connections constant for each node and link type. This holds the marginals roughly constant but rewires the network topology. Then, the same procedure as for the empirical data was used to obtain $Q(q)_{rnd}/Q_{max,rnd}$ which is also averaged over several realizations of the disorder.

In the left part of Fig. 7.1 we compare the values of $Q(q)/Q_{max}$ for the empirical data and the randomized data. While the randomized data show a linear increase with the number of roles from the beginning, the empirical data show a strong increase at small numbers of roles and then also changes into a linear regime.

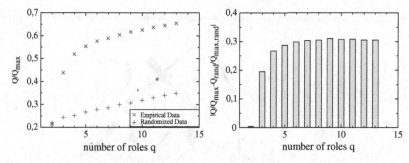

Fig. 7.1. *Left*: Average of $Q(q)/Q_{max}$ over five commodities for the world trade network as a function of the number of roles q in the block model. Red (x) denotes the actual empirical data, blue (+) denotes the results averaged over randomly rewired versions of the empirical data as a null model. While the randomized data show a linear increase of Q/Q_{max} with the number of roles, the empirical data exhibit a strong increase for smaller numbers of q and then also turns into a linear regime. *Right*: Difference between Q/Q_{max} for the empirical data and the randomized data. At $q = 5$ we observe the transition to the linear regime. At $q = 9$ the largest difference between empirical data and the random null model occurs capturing 60% of Q_{max} with only 8% of the total number of structural equivalence classes needed to achieve this maximum.

The right part of Fig. 7.1 shows the difference in the ratio $Q(q)/Q_{max}$ of empirical and randomized data. Though every block model from $q = 2$ to $q = 112$ has its own merit, after all, the countries do all have individuality, two points may be chosen as particularly meaningful: either the number of roles at which we observe the transition to a linear increase in $Q(q)/Q_{max}$ which happens at $q = 5$ or the point at which we observe the largest difference to the randomized data at $q = 9$. An alternative approach to select the optimal number of roles would be to use the minimum description length of the block model as suggested in Ref. [4].

Note that as the number of roles increases, their memberships may merge as well as split. Successive partitions are not always subdivisions forming hierarchical clusters, although there is a strong tendency for that to occur. The five rectangles enclosing pairs of roles in Fig. 7.2 show where subdivisions tend to be hierarchical. In each case, however, some other countries also join the new sub-roles, as, for example, when the less-developed periphery of the two-role model splits into two sub-roles that are also joined by some countries from the core.

Figure 7.3 shows the image graphs and block matrix plots for five and nine roles. Note the progression of differentiation as more and more roles are included. Already at $q = 5$ we observe a structure that can be seen as a coarse-grained version of the model with nine roles, with the models in between mediating the transition. Inspection of Table 7.1 shows for all the block models that the progressive refinements in Fig. 7.3 induce a fair approximation to a

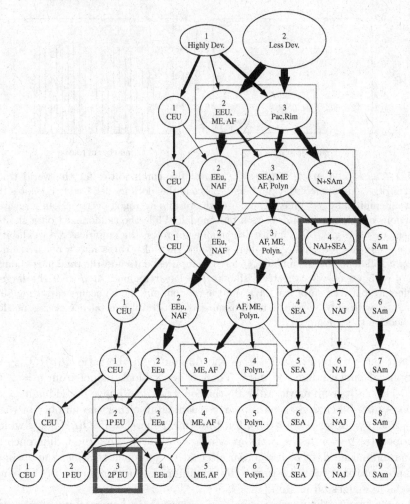

Fig. 7.2. Splitting and merging diagram of the assortment of countries into roles as the number of roles increases. The width of the arrows is proportional to the number of countries that pass from role to role as the number of classes is increased by one. Thin line rectangles indicate the major split on each level and thick line rectangles show new roles formed from overlap or merging. See Table 7.1 for the individual countries in each role at each level. The compact layout of this splitting/merging diagram shows how splits tend to distribute countries to smaller blocks that are adjacent in the partition order. This suggested the compact order of the blocks in Table 7.1 and Fig. 7.3. The only three exceptions to compactness are China's realignment to block 1 at level 3 and back to block 4 at level 4 and Saudi Arabia's realignment to block 3 at level 5. As already noted in the matrix plots and image graphs, differentiation first happens around the Pacific and then in Europe, Africa and the Middle East. Labels are CEU: Central Europe, EEU: Eastern Europe, ME: Middle East, AF: Africa, NAF: Northern Africa, SEA: South East Asia, Polyn: Polynesia, N+SAm: North and South and Middle America, SAm: South and Middle America, NAJ: North America and Japan, 1P EU: 1st periphery EU, 2nd periphery EU.

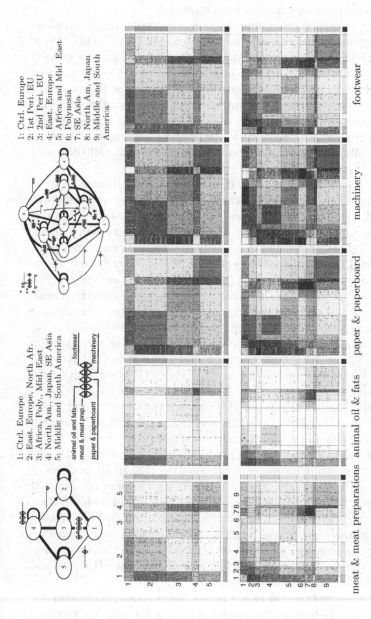

Fig. 7.3. Consensus image graphs and block matrix plots for the five commodities studied at $q = 5$ and $q = 9$ roles. Note the high symmetry of the image graphs. Triangle labels indicate commodity and direction of the flow of goods. Unlabeled links carry all five commodities in both directions. Side and bottom bars encode the marginal fraction of import and export of the total traded volume for each block in gray scale, respectively. Black dots indicate trade greater than expected from the marginals for *pairs* of countries, white dots smaller than expected. Background shading of blocks corresponds to density of black dots in block. See Table 7.1 for individual countries grouped in each block and text for details. The ordering of blocks in the matrices is suggested by the proximity order of the splitting diagram as depicted in Fig. 7.2.

hierarchical clustering of roles. This is not required by the model and rather than split as the number of roles increases, memberships merge from different blocks about 8% of the time, including cases where two roles keep their identity but contribute overlapping members to form a third.

A pattern of geographical proximities appears in an ordering of partitions that minimizes distances between sets that are merged or split. This unique compact layout of the splitting/merging diagram (comp. Fig. 7.2) is used to order the partitions in Table 7.1. Countries in the same group tend to be located in each other's proximity. Geographical position is thus a strong factor determining the grouping of countries. One reason is of course that geographical proximity means that such countries have similar geographical conditions and hence similar conditions for agriculture, mining, etc. Another is that geographically close countries often form localized trade alliances and trade is facilitated by short distances.

Additional to geographic proximity, the second striking feature of these block models is that there exists considerable symmetry in the way the world trade is organized. Symmetry of the image graphs suggests that there are also regular equivalences across regions that organize the role structures across different regions [3]. Let us consider the $q = 9$ model. Thus, on one hand, there is the region around the Pacific with the United States, Canada and Japan (8) in a central position, South America (9) as an out-group and South East Asia (7) as a sub-center. On the other hand, we see the core of the European Union (1) in an equally central position as the United States, Canada and Japan (8), however, with Eastern Europe and the former Soviet Union (4) assuming the position that South America (9) takes on across the Atlantic. Scandinavia and some peripheral European countries such as Ireland, Austria, Greece and also Turkey (2) are for the core EU states (1) what South East Asia (7) is for North America and Japan (8). In the middle of all, we find the African and Middle Eastern countries (5), Polynesia (6) and a second group of peripheral European countries (3) which are Greenland, Iceland, Portugal, Andorra, Malta and Israel in approximately equal positions.

It is also interesting to observe the gradual refinement of the roles, for instance when concentrating on the core EU countries. For small numbers of roles, countries such as Denmark, Sweden, Austria and Norway are grouped together with them, but with more roles available, they are moved into their own groups to merge with countries such as Cyprus, Finland and Ireland which had been in more peripheral positions from the start. Such behavior can be interpreted as showing, with greater refinement in the role structure, the intermediary positions between the clear role of the core EU states and the more peripheral countries.

Our choice of penalty function (3.22) makes the proposed framework for block modeling a density-based measure but not, as in some earlier methods [5], based on a notion of merely high/low densities within position-to-position blocks compared to *global* densities. Rather, its partitions are based on the marginal expectations from the paired row–column positional totals that meet in a given block, i.e., where links are concentrated. We thus take a different approach than the parameter-rich mixture model approaches in Refs. [4,6–10]. The method allows the use of weighted data sets of multiple link types which is difficult for block modeling procedures based on probabilistic approaches.

Successive partitions are not necessarily sequential hierarchical sub-clusters but may be overlapping. This allows for modeling the fact that actors do not usually take on a single role but an intersection of roles. The proposed framework may help to recover some of these intersections through the overlapping partitions that occur with different granularities of roles.

In conclusion, this first application has shown that the approach presented is able to recover meaningful assignments of nodes in a network into classes of what one may call structural similarity. The choice of penalty function and hence expectation values based on the marginals of the country to country trade matrix has led to a grouping of countries mainly according to geography. The links in the image graphs hence represent trade routes where the volume exceeds the expectations. As such, the image graphs differ largely from other maps of world trade which often emphasize the high volume links such as the connection between the European Union and North America. It should be noted that other choices of penalty functions may lead to very different image graphs and groupings of countries into roles. In this lies the great flexibility of the presented approach to block modeling, as it allows us to discover role structures which emphasize a number of different aspects of the network structure.

Table 7.1. Assignment of countries in models with two to nine roles. The horizontal lines separate the $q = 9$ different roles of the most detailed block model from Fig. 7.3. Note how the blocks form an almost perfect hierarchy in the way that successive blocks split apart although this is *not* required by the algorithm. This is also shown by the splitting diagram in Fig. 7.2 which further suggests the order of the groups of countries in this table.

Group label	Country	q=2	q=3	q=4	q=5	q=6	q=7	q=8	q=9
	Belgium-Luxembourg	1	1	1	1	1	1	1	1
	France	1	1	1	1	1	1	1	1
	Germany	1	1	1	1	1	1	1	1
	Italy	1	1	1	1	1	1	1	1
Core EU	Netherlands	1	1	1	1	1	1	1	1
	Spain	1	1	1	1	1	1	1	1
	Switzerland	1	1	1	1	1	1	1	1
	United Kingdom	1	1	1	1	1	1	1	1
	Denmark	1	1	1	1	1	1	2	2
	Sweden	1	1	1	1	1	1	2	2
	Austria	1	1	2	2	2	2	2	2
1st Peri. EU	Turkey	1	2	2	2	2	1	2	2
	Greece	1	2	2	2	2	2	2	2
	Norway	1	2	2	2	2	2	2	2
	Finland	2	2	2	2	2	2	3	2
	Ireland	2	2	2	2	2	2	2	2
	Cyprus	2	2	2	2	2	2	2	2
	Portugal	2	2	2	2	2	2	2	3
	Andorra	2	2	2	2	2	2	3	3
2nd Peri. EU	Iceland	2	2	2	2	2	2	3	3
	Israel	2	2	2	2	2	2	3	3
	Greenland	2	2	3	2	2	2	3	3
	Malta	2	2	2	2	2	2	4	3
	Russian Federation	1	1	2	2	2	2	2	4
	Czech Rep.	1	2	2	2	2	2	2	4
	Turkmenistan	1	2	2	2	2	2	3	4
	Albania	2	2	2	2	2	2	3	4
	Armenia	2	2	2	2	2	2	3	4
	Azerbaijan	2	2	2	2	2	2	3	4
	Belarus	2	2	2	2	2	2	3	4
	Bulgaria	2	2	2	2	2	2	3	4

Table 7.1. (Continued)

Group Label	Country	q=2	q=3	q=4	q=5	q=6	q=7	q=8	q=9
East. Europe	Georgia	2	2	2	2	2	2	3	4
	Hungary	2	2	2	2	2	2	3	4
	Iran	2	2	2	2	2	2	3	4
	Kazakhstan	2	2	2	2	2	2	3	4
	Latvia	2	2	2	2	2	2	3	4
	Lithuania	2	2	2	2	2	2	3	4
	Poland	2	2	2	2	2	2	3	4
	Rep. of Moldova	2	2	2	2	2	2	3	4
	Romania	2	2	2	2	2	2	3	4
	Serbia and Montenegro	2	2	2	2	2	2	3	4
	Slovakia	2	2	2	2	2	2	3	4
	Tajikistan	2	2	2	2	2	2	3	4
	Syria	2	2	2	2	2	3	4	4
	Saudi Arabia	1	1	1	1	3	3	4	5
	Algeria	2	2	2	2	2	3	4	5
	Morocco	2	2	2	2	2	3	4	5
	Tunisia	2	2	2	2	2	3	4	5
	Bahrain	2	2	3	3	3	3	4	5
	Comoros	2	2	3	3	3	3	4	5
	Cote d'Ivoire	2	2	3	3	3	3	4	5
	Ethiopia	2	2	3	3	3	3	4	5
	Ghana	2	2	3	3	3	3	4	5
Africa, Mid. East	Guinea	2	2	3	3	3	3	4	5
	Jordan	2	2	3	3	3	3	4	5
	Nigeria	2	2	3	3	3	3	4	5
	Oman	2	2	3	3	3	3	4	5
	Senegal	2	2	3	3	3	3	4	5
	Togo	2	2	3	3	3	3	4	5
	Burundi	2	3	3	3	3	3	4	5
	Kenya	2	3	3	3	3	3	4	5
	Mauritius	2	3	3	3	3	3	4	5
	Pakistan	2	3	3	3	3	3	4	5
	Uganda	2	3	3	3	3	3	4	5
	China, Macao SAR	2	3	3	3	3	4	5	6
	French Polynesia	2	3	3	3	3	4	5	6
	Maldives	2	3	3	3	3	4	5	6
	Nepal	2	3	3	3	3	4	5	6
Polynesia	New Caledonia	2	3	3	3	3	4	5	6
	New Zealand	2	3	3	3	3	4	5	6
	Papua New Guinea	2	3	3	3	3	4	5	6
	Philippines	2	3	3	3	3	4	5	6
	Vanuatu	2	3	3	3	3	4	5	6
	Malaysia	2	3	3	3	3	5	6	7
	Indonesia	2	3	3	3	4	5	6	7
	Singapore	2	3	3	3	4	5	6	7
	South Africa	1	3	3	3	4	5	6	7
SE Asia	Thailand	1	3	3	3	4	5	6	7
	Australia	1	3	3	4	4	5	6	7
	China	1	3	1	4	4	5	6	7
	China, Hong Kong SAR	1	3	3	4	4	5	6	7
	Rep. of Korea	1	3	3	4	4	5	6	7
	Japan	1	3	3	4	5	6	7	8
North Am, Japan	Canada	1	3	4	4	5	6	7	8
	USA	1	3	4	4	5	6	7	8
	Brazil	1	3	4	4	6	7	8	9
	Argentina	1	3	4	5	6	7	8	9
	Barbados	1	3	4	5	6	7	8	9
	Honduras	1	3	4	5	6	7	8	9
	Panama	1	3	4	5	6	7	8	9
	Bolivia	2	3	4	5	6	7	8	9
	Chile	2	3	4	5	6	7	8	9
	Colombia	2	3	4	5	6	7	8	9
	Costa Rica	2	3	4	5	6	7	8	9
	Dominica	2	3	4	5	6	7	8	9
	Ecuador	2	3	4	5	6	7	8	9
South America	El Salvador	2	3	4	5	6	7	8	9
	Guatemala	2	3	4	5	6	7	8	9
	Jamaica	2	3	4	5	6	7	8	9
	Mexico	2	3	4	5	6	7	8	9
	Montserrat	2	3	4	5	6	7	8	9
	Nicaragua	2	3	4	5	6	7	8	9
	Paraguay	2	3	4	5	6	7	8	9
	Peru	2	3	4	5	6	7	8	9
	St Kitts and Nevis	2	3	4	5	6	7	8	9
	St Lucia	2	3	4	5	6	7	8	9
	St Vincent & Grnads.	2	3	4	5	6	7	8	9
	Suriname	2	3	4	5	6	7	8	9
	Trinidad and Tobago	2	3	4	5	6	7	8	9
	Uruguay	2	3	4	5	6	7	8	9
	Venezuela	2	3	4	5	6	7	8	9

7.2 Communities of Common Interest among eBay Users

Our second example will also be a trade network. However, this time we will
deal with a network of individual persons which interact by competing on a
large market. Instead of looking for the best fitting block model, this time, we
will try to find the optimal fit of the network to a diagonal block model which
corresponds to a grouping of agents into communities of common interest.

The Internet has changed the way people communicate, work and do business. One example is online auction sites, the largest being eBay with its more than 150 million registered users worldwide [11]. An interesting aspect of eBay's success is its transparency. The market is fully transparent as the trading history of every user is disclosed to everyone on the Internet. In this chapter, a network of competitors in this market is investigated and users are clustered into groups with homogeneous buying interest profiles, so-called market segments. In economics, market segmentation studies are often the starting point in designing targeted marketing campaigns and the quality and validity of this exploratory data analysis are crucial for their success. However, such studies often have to deal with several difficulties: First, a specific similarity measure has to be tailored and second, if the data are very high dimensional and sparse it requires dimensionality reduction to make conventional methods applicable. This, however, may introduce bias in the analysis. In contrast, our method of block modeling is able to work directly on the raw data and without the introduction of any similarity measure. Our results show how market participants use the online auction site and where there is economic growth potential for the host apart from increasing the number of users of the site.

Let us first recall the operating principle of an online auction in Fig. 7.4. Users may offer goods through the online platform and set a deadline when their auction will end. Articles are listed under a certain taxonomic product category by the seller and are searchable platform wide. Users with a particular demand either browse through the articles listed in an appropriate category or search for articles directly. Until the end of the auction they may bid on the article. The user with the highest bid at the end of the auction wins (so-called hard-close) and buys the article. In every new auction, users may assume different new roles as sellers, bidders or buyers. The market can

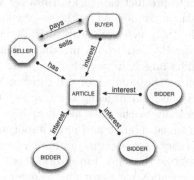

Fig. 7.4. Structure of a single auction. Users express their common interest in a particular article by bidding. The user with the highest bid wins the auction and exchanges money and the article with the seller. eBay earns a fee with every transaction. Users of the auction site, i.e., bidders, buyers or sellers, may change their role in a different auction of another article.

be represented as a graph with the users and/or articles as the nodes and the links denoting their interactions as shown in Fig. 7.4.

A number of researches have presented statistical studies of trading [12] and analyses of bidding strategies and auction ending rules [13, 14]. Here, the focus lies on the market segmentation of the eBay auction site. At a certain level of abstraction the population of consumers can be assumed to be separated into relatively clear-cut and homogenous subgroups corresponding to certain customer milieus or market segments [5]. Customers of the same type are described by a common pattern in their consumer interests which leads to a higher probability of bidding for the same article [15] and thus to a higher density of interactions between users of the same type. This is the reason we will directly try to fit a diagonal block model to the network. We will solely use the information of which users competed in the same auctions, i.e., for single articles. This fitting of a diagonal block model can be interpreted as a kind of cluster analysis [16, 17] of the bidding behavior of the users in our data set. The classification is possible even for this very sparse and high-dimensional data set [18] with each bidder on average taking part in only slightly more than 3 out of 1.6 million possible auctions.

Classical clustering techniques generally fail for such high-dimensional data due to the "curse of dimensionality" [19], a problem which arises when the dimension D of the data set to be clustered increases [18]. The data points become increasingly sparse as the dimensionality increases and the relative difference between the closest and the farthest neighbor of a randomly selected point in the data set goes to zero with increasing dimension [18, 20]. In our case, the dimension of the space of articles is 1.6 million.

Other conventional analysis techniques such as correspondence analysis [5, 21] have to make use of a similarity measure between articles in order to reduce the dimensionality and coarse-grain the data, such as exploiting the annotation of articles into product categories. However, this bears several pitfalls: First, the annotations are defined by the seller who lists the article such that it can be found efficiently, hence, the categorization is mainly a taxonomy. Using this to coarse-grain the data would introduce a bias in the analysis. Second, eBay categories differ largely in size when counting the number of articles in the category as well as the number of sub-categories. Correcting for this again may introduce a bias. Third, using the category taxonomy for coarse-graining induces a hierarchy in the data, as all articles below the cut in the taxonomy tree are subsumed. Fourth and most importantly, it is not clear at which level in the category tree a coarse-graining should be performed and whether this level should be the same for all branches.

Our analysis is independent of taxonomic categories and dimensionality reduction. We will show how to find evidence for hierarchical and overlapping cluster structures as well. The product categories are solely used to interpret the results of the study, i.e., provide interest profiles of user groups found in terms of this taxonomy.

By clustering users directly according to a common demand spectrum, problems of conventional basket analysis done by frequent item sets [22–25] are also circumvented. The latter asks which articles are frequently demanded by a single person. This analysis is performed for all articles averaging over the entire population of consumers and hence results in the least common denominator of articles which may then be bundled together and marketed together to the whole population of customers. The same is true for cluster analysis of eBay categories [14]. The block modeling procedure used here, however, reveals information about people and their diverse and possibly very special interests.

7.2.1 Data Set

A data set consisting of over 1.59 million auctions ending during the pre-Christmas season December 6–20, 2004. Considering only articles located in Germany, the user id of seller, buyer and all bidders competing in each auction was recorded, as well as the individual bids and the product category in which the article was listed (excluding articles listed in the real estate category which was in a beta testing phase at the time). Since auctions last between 7 and 10 days depending on the choice of the seller, a bidding period of up to 25 days is covered was obtained from the German eBay site www.ebay.de.

The pre-Christmas time is a suitable time for analysis for the following reasons: First, traffic is very high. In fact, there was a broad advertising campaign in Germany advertising to shop for Christmas presents on eBay. Second, only auctions are considered and one expects that users are unlikely to bid for articles for which they cannot assess a fair price. Third, if users shop for presents, then one can gain some information about their family background, e.g., people shopping for toys will most likely have a child themselves or among their closer relatives. The results indicate that this is indeed the case. Table 7.2 summarizes the data set in its basic parameters. There are far less sellers than bidders and only 38% of the sellers also act as bidders or buyers. This indicates that users are split into those mostly selling and those mostly buying.

Table 7.2. Summary of the data set of online auctions obtained between December 6 and 20, 2004. Numbers in millions.

Auctions observed:	1.59
Users acting as buyer:	0.95
Users acting as seller:	0.37
Users acting as bidder:	1.91
Users acting as seller and bidder:	0.14
Users acting as seller and buyers:	0.08

7.2.2 User Activity

The activity of the users is measured via the probability mass distributions of the number of articles sold $p(s)$, bought (auctions won) $p(w)$ and bid on $p(a)$. Though it is possible to bid multiply in a single auction, we neglect this fact and use "bid" and "take part in an auction" synonymously. Similar to previous studies [12], we find fat-tailed distributions of the user activity. Due to the short time span observed and a constant growth of the market, one cannot regard these distributions as representing a steady state. Nevertheless, some insight can be obtained. We compared maximum likelihood fits of the data to log-normal distributions of the form $p(x) \propto (x - \theta)^{-1} \exp[-\frac{1}{2}(\ln((x - \theta)/m)/\sigma)^2]$ and power laws of the form $p(x) \propto (x + \Delta x)^{-\kappa}$ and find that both kinds of distribution characterize the data almost equally well with a slight advantage for the power law especially for the very rare events in the tail of the distribution as can also be observed in the cumulative plots. In Ref. [12], only power laws were considered. For the number of bidders b taking part in an auction, the "attractiveness of an article", we consider an exponential distribution $q(b) \propto \alpha^b$. A possible alternative distribution, the binomial or Poisson distribution can be ruled out, as the empirical distribution is monotonously decreasing and these distributions would require a maximum at the average value. Alternatively, an almost perfect fit can be achieved when assuming $q(b) \propto \exp(-b^{1.117}/3.7)$. However, we prefer the simple model with only one free parameter $\alpha = 1 - 1/\langle b \rangle$ and attribute the observed deviation at high numbers of bidders per article to a saturation effect. We believe that if an article has attracted a critical number of bidders, potential additional bidders are more reluctant to join because of the already strong competition and hence there are up to 10 times less articles with 15 or more bidders than expected from the simple model. Recalling that only 1 in 1000 auctions attracts more than 15 bidders, naturally, our hypothesis would have to be confirmed or rejected by future research. Figure 7.5 shows a graphical representation of these distributions and Table 7.3 summarizes the parameters obtained by maximum likelihood fitting [26, 27].

The distribution of the number of articles sold per seller falls off slowest, followed by the number of articles bid on and the number of articles bought. Here, we see the professionalization on the seller side of the market. There are "power-sellers" making a living from selling via eBay, but there are hardly any "power-buyers" professionally buying on eBay. This shows that eBay is more of a selling platform than an actual trading site, where selling and buying activities would be more balanced.

If we assume that the tail of the distribution of the number of articles sold per seller is representative of the "firm size" of these users and compare these to the long-term statistics of firm sizes in the United States given by Axtell [28], we can confirm the power law tail of the distribution, but not the exponent of $\kappa = 2$. Instead, we find $\kappa = 2.31$ and thus the observed distribution falls off faster. We can only speculate on the reasons for this and

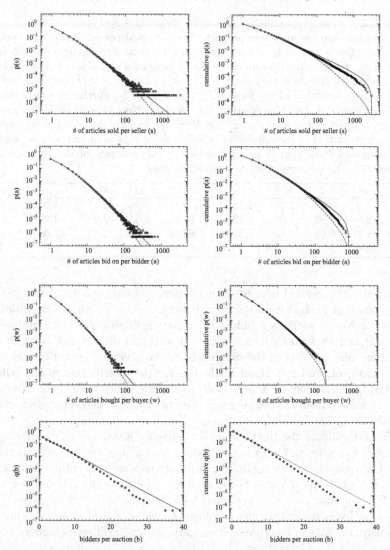

Fig. 7.5. User activity during the pre-Christmas season 2004. From *top left* to *bottom right*: probability mass function (*left*) and cumulative probability distribution (*right*) of the number of articles sold $p(s)$, different auctions participated in $p(a)$ and number of articles bought $p(w)$ as well as number of bidders participating in an auction $q(b)$. For the first three distributions, the red solid lines correspond to maximum likelihood fits with power laws $p(x) \propto (x + \Delta x)^{-\kappa}$ while the blue dashed lines correspond to fits with a log-normal distribution of the form $p(x) \propto (x - \theta)^{-1} \exp[-\frac{1}{2}(\ln((x - \theta)/m)/\sigma)^2]$. The fits for the cumulative distribution take the finite size of the sample into account. For the distribution of the desirability of an article $q(b)$, the red line represents a maximum likelihood exponential fit of the form $q(b) \propto \alpha^b$. Parameter estimates and log likelihood scores can be found in Table 7.3.

Table 7.3. Estimated parameters of activity distributions of observed users in auctions. For the distribution of the number of articles sold per seller $p(s)$, the number of auctions taken part in by a bidder $p(a)$ and the number of articles bought by a buyer $p(w)$, we show the maximum likelihood estimates of the parameters for both a log-normal distribution $p(x) \propto (x - \theta)^{-1} \exp[-\frac{1}{2}(\ln((x - \theta)/m)/\sigma)^2]$ as well as for a shifted power law of the form $p(x) \propto (x + \Delta x)^{-\kappa}$. Further, we indicate the likelihood scores of these distributions for the data. Note that all distributions are better described as power laws, but the difference to the log-normal is very small. For the distribution of the number of bidders taking part in an auction, we assume an exponential form $q(x) \propto \alpha^x$ and show the ML estimate of the parameter α. Additionally, the averages of all quantities are given.

$p(x)$	$\langle x \rangle$	θ	m	σ	L_{LN}	Δx	κ	L_{Pow}	α
$p(s)$	4.3	0.93	1.28	1.55	−1.9427	0.96	2.31	−1.9419	
$p(a)$	2.9	0.82	1.08	1.32	−1.6722	1.55	2.90	−1.6719	
$p(w)$	1.7	0.90	0.60	1.21	−1.0415	0.90	3.43	−1.0414	
$q(b)$	3.4								0.71

further study is needed here to compare new and old economy. In an earlier study, Yang et al. had reported an exponent of $\kappa = 3.5$ for the distribution of the number of auctions a bidder takes part in from a data set obtained in 2001 [12] and we found $\kappa = 2.9$ in our data. If this discrepancy is the result of a trend and not due to the differences in the observed countries and sizes of the data set, and this trend holds also for the distribution of the seller's activity, then one may be able to observe a convergence toward the exponent of $\kappa = 2$ known from the old economy. Further study is needed regarding this hypothesis.

The fat tails of the distribution are striking given the short time span observed. Consider the most active bidder taking part in over 800 auctions! This user seems to follow a gambling strategy bidding only minimal amounts as he/she wins only a few of these auctions. The most successful buyer who won 201 auctions on the other hand took part in only 208 auctions. This hints at a diversity of strategies employed by users of the online auction site. Curiously, the article most desired and attracting 39 different bidders was a ride in a red Coca-Cola-Truck. The fat tails of the distribution also show that there is no "typical" user activity, rather, one observes activities at all scales.

7.2.3 User Networks

From the original data a number of market networks can be constructed. The most natural one would of course be the network of users connected by actual transactions. Another would be the network of sellers that are connected if they have sold to the same user. Then, the links in the network would represent a possible competition or a possibility for cooperation, depending on the portfolio of articles offered by these sellers. This situation is also known as "co-opetition". Similarly, such a co-opetition network could also be constructed based on the fact that different sellers have received bids from the same user. Further, the construction of customer groups based on the fact that they have bought from the same seller is possible which again would define a co-opetition situation for the sellers that join these customers. One could also study the relations of articles or categories based on joining them to networks when they have received bids from the same users. This would then resemble a frequent item or frequent category analysis.

Here, the focus lies on the bidder network based on single articles. Two bidders are linked if they have competed in an auction. Since all users who bid in a single auction are connected, this network results from overlaying fully connected cliques of bidders that result from each auction. Such graphs are also known as affiliation networks [29–31]. Note that one could also assign weights to links between bidders according to the number of times they have met or according to some function of the amount of money they have bid.

Prior to a block modeling analysis in this bidder network, we study its general statistical properties looking for indications of block structure [32]. We compare the results to a randomized null model (RNM) obtained from reshuffling the original data, i.e., keeping the attractiveness of each auction and the activity of each bidder constant, but randomizing which bidders take part in which auction. If the presence of clusters of users with a common interest has an influence on the statistical parameters of the network, it should be detectable by comparison with such a random null model. Figure 7.6 shows a comparison between the empirical data and the RNM in terms of the cumulative degree distribution, the cumulative distribution of the link weights as well as the clustering coefficient $c(k)$ as a function of the degree k.

The shapes of the cumulative degree distributions agree quite well between the RNM and the original data. However, for higher values of k, the distribution of the empirical data lies below that of the RNM. Since meeting the same competitor twice in different auctions does not lead to an increase in the number of neighbors in the network, this shows that competitors meet more often in the real world than expected from the random null model.

Furthermore, the bidder network is compared with theoretical predictions derived from the distribution of the bidding activity and the distribution of the number of bidders per auction which can be obtained by employing the generating function formalism [31]. Two generating functions are defined for the distribution p_a of the number of different auctions a a bidder takes part in and q_b for the number of different bidders b that compete in an auction [31]:

$$f_0(x) = \sum_{a=0}^{\infty} p_a x^a \text{ and } g_0(x) = \sum_{b=0}^{\infty} q_b x^b. \tag{7.2}$$

It is assumed that $p_0 = q_0 = 0$ in derivations hereafter, since the data set contains only bidders that have taken part in at least one auction and only those auctions which had attracted at least one bidder were recorded. The

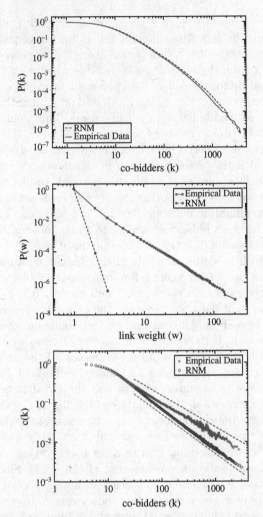

Fig. 7.6. Comparison of the bidder network with a random null model (RNM).
Top: Cumulative degree distribution $P(k)$. The degree distribution found empirically (solid red line) lies below that of the random null model (dashed blue line).
Middle: Cumulative distribution of the link weights in the bidder network. The distribution for the empirical data (solid red line) lies above that of the random null model (dashed blue line). *Bottom:* Distribution of the clustering coefficient $c(k)$ as a function of the degree k of the nodes. The clustering is higher for the empirical data (red crosses) than for the RNM (blue circles). The two dashed lines indicate two power laws $\propto k^{-\kappa}$ with exponents $\kappa = 1$ and $\kappa = 0.8$, respectively, for comparison.

mean number of auctions a bidder takes part in and the mean number of bidders per auction are then given by

$$\langle a \rangle = f_0'(x = 1) \text{ and} \tag{7.3}$$

$$\langle b \rangle = g_0'(x = 1). \tag{7.4}$$

The specific forms of p_a and q_b are taken from the empirical data. The exponential distribution for the attractiveness of an auction for instance suggests $q_b = (1-\alpha)\alpha^{b-1}$ from which $\langle b \rangle = 1/(1-\alpha)$ follows, which is just the equation by which the parameter α is estimated using the maximum likelihood method. The degree distribution of the bidder network can be calculated from p_a and q_b using a generating function formalism. Assuming that bidders never meet twice in different auctions, the generating function of the degree distribution in the bidder network is [31]

$$G_0(x) = f_0(g_0'(x)/g_0'(1)). \tag{7.5}$$

The functions $f_0(x)$ and $g_0(x)$ are the generating functions for the bidding activity and the attractiveness of an article as introduced by (7.2). The degree distribution then follows from the derivatives of the generating function with respect to x:

$$p(k) = \frac{1}{k!}\frac{d^k}{x^k}G_0(x)|_{x=0} \tag{7.6}$$

$$= \alpha^k \sum_a p_a (1-\alpha)^{2a} \binom{2a-1+k}{k}. \tag{7.7}$$

A theoretical expectation for the average number of neighbors in the bidder network can be estimated from the average number of bidders per auction $\langle b \rangle$ and the average number of auctions a user takes part in $\langle a \rangle$:

$$\langle k \rangle = G_0'(1) = \langle a \rangle \frac{\langle b^2 \rangle - \langle b \rangle}{\langle b \rangle} = \langle a \rangle \frac{2\alpha}{1-\alpha} \tag{7.8}$$

$$= 2(\langle b \rangle - 1)\langle a \rangle. \tag{7.9}$$

This yields an estimated value for the average number of links in the network of $\langle k \rangle = 14$ which is in excellent agreement with the result from the RNM ($\langle k \rangle = 13.9$), but larger than in the actual data ($\langle k \rangle = 12.9$) confirming our expectation. See Table 7.4 for a summary of the basic parameters of the empirical data and the RNM.

Comparing the cumulative distribution of the link weights, i.e., the number of times two bidders have met in different auctions, we find a much more prominent difference between the data and the RNM. Figure 7.6 shows that the weights of the links in the bidder network are distributed with a power law tail. Approximately 6% of all links correspond to pairs of bidders who have met more than once. If there would be no common interest among bidders, practically all links would have weight 1 as is indeed the case for the RNM.

Table 7.4. Summary of basic parameters for the bidder network with two bidders linked, if they have competed in an auction. Shown are the actual data, the parameters for a random null model (RNM) obtained by reshuffling the bidders in different auctions and the reduced version of the network used for cluster analysis containing only those bidders having taken part in more than one auction.

	Data	RNM	Reduced
Number of nodes:	1.8×10^6	1.8×10^6	0.9×10^6
Number of links:	11.6×10^6	12.6×10^6	7.4×10^6
Average degree:	12.9	13.9	16.4
Assortativity:	0.02	$-(4 \pm 3) \times 10^{-4}$	0.03

Additionally to the distribution of degrees and link weights, we compare the distribution of the clustering coefficient as a function of the degree of a node. The clustering coefficient $c(k)$ denotes the average link density among the neighbors of a node of degree k. Due to the construction process of the network as an affiliation network, we expect that for large numbers of neighbors k the clustering coefficient $c(k)$ scales as k^{-1} in the case of random assignment of bidders to auctions [30]. Figure 7.6 shows that this is indeed the case for the RNM, but the actual data deviate strongly for bidders with a large number of neighbors and show higher clustering. This effect can arise from two processes: either bidders with whom one competes in two different auctions also meet independently in a third auction or that there is an increased probability that one will compete again with a bidder one has already met once in an auction. Both explanations support our assumption of the presence of clusters of users with common interest. Note that the scaling of the correlation coefficient with the degree of the nodes and exponent -1 is purely a consequence of the construction process of the network and not an indication for hierarchical modularity as introduced by Ravasz et al. [33].

With these comparisons, the bidder network is shown to be far from randomly constructed and one can proceed by studying the block structure for which indirect evidence was found already. Table 7.4 summarizes again the basic parameters of the bidder network, the randomized null model and the reduced version of the bidder network which will be used for fitting a diagonal block model in the following section.

7.2.4 Market Segmentation

7.2.4.1 Network Clustering

The analysis of the user interests in the eBay market is based on the bidder network as constructed in the previous section. The links in this network represent articles the connected bidders (nodes) have a common interest in. The network is reduced to only those bidders that have taken part in at least two auctions and only auctions with a final price below $1,000$ Euro are considered, thereby focussing on consumer goods. See Table 7.4 for the basic parameters of this reduced network.

If one now finds groups of users (clusters or communities [34–36]) with a high density of links among themselves and a low density of links to the rest of the network, the total set of links within such a group of users can be interpreted as a unifying common interest of this group. We fit a diagonal block model as introduced in Chap. 4. Recall the quality function Q used:

$$Q = \sum_s \underbrace{(m_{ss} - \gamma[m_{ss}])}_{c_{ss}} = -\sum_{s<r} \underbrace{(m_{rs} - \gamma[m_{rs}])}_{a_{rs}}. \tag{7.10}$$

Note that any assignment of bidders into groups which maximizes Q will be characterized by both maximum cohesion of groups and minimal adhesion between groups. If Q is maximal, every node is classified in that group to which it has the largest adhesion. Compare Chap. 4 again for examples and further details of this quality function. Maximally 500 different groups of bidders were allowed in the analysis which gives a sufficient level of detail.

Figure 7.7 compares the results obtained with $\gamma = 0.5$ and $\gamma = 1$. Shown are the adjacency matrices A_{ij} of the largest connected component of the bidder network. A black pixel at position (i,j) and (j,i) is shown on an $889,828 \times 889,828$ square if bidders i and j have competed in an auction and hence $A_{ij} = 1$, otherwise the pixel is left white corresponding to $A_{ij} = 0$. The rows and columns are ordered such that bidders who are classified as being in the same group are next to each other. The internal order of bidders within groups is random. The order of the groups was chosen to optimally show the correspondence between the ordering resulting from the $\gamma = 0.5$ and the $\gamma = 1$ ordering. In this representation, link densities correspond to pixel densities and thus to gray levels in the figure. Information about the exact size and link density contrast of the clusters is given in Table 7.5. Note the high contrast between internal and external link density.

At the top of Fig. 7.7, the adjacency matrix ordered according to an optimal assignment of bidders into groups with $\gamma = 0.5$ is shown. Clearly, a small number of major clusters of bidders and a large number of smaller clusters are identified, strongly connected internally and well separated from one another. The largest eight clusters are marked with letters A through H. Of all bidders in the network, 85% are classified in these eight clusters. At the bottom, the same adjacency matrix is shown, but now rows and columns are ordered according to an optimal assignment of bidders into groups with $\gamma = 1$.

As expected, a larger number of smaller, denser clusters are found, which are numbered 1 through 13. In order to analyze whether the network has a hierarchical or overlapping cluster structure, a consensus ordering of the bidders from the $\gamma = 0.5$ and $\gamma = 1$ ordering is defined by reshuffling the internal order of the $\gamma = 0.5$ clusters according to the $\gamma = 1$ clustering. Remember the orderings for the two values of γ were obtained independently.

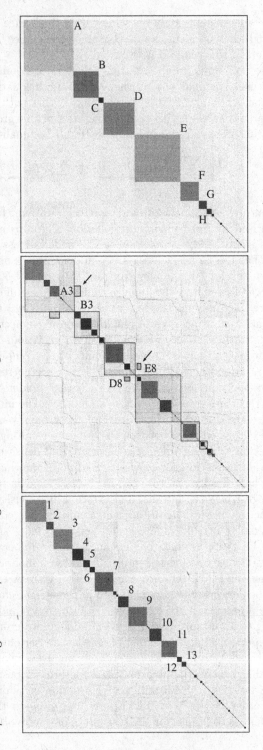

Fig. 7.7. N×N adjacency matrix of the bidder network in three different orderings. A pixel in row i, column j corresponds to an auction in which bidders i and j have competed. Shown are $N = 889,828$ bidders (nodes) and $M = 7,373,008$ pairwise competitions (links). Gray levels correspond directly to link density in this network and hence to the probability of competing in an auction. *Top:* $\gamma = 0.5$ ordering. *Bottom:* $\gamma = 1$ ordering. *Middle:* consensus ordering of top and bottom. See text for details.

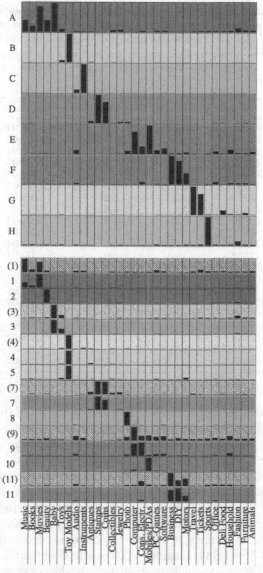

Fig. 7.8. Risk ratios of bidding in one of the 32 main eBay product categories for classified users. *Top*: from $\gamma = 0.5$ classification. *Bottom*: from $\gamma = 1$ classification. Spectra with a dashed background (cluster id in parenthesis) show customer purchases $6-9$ months after original classification. See text for details.

Table 7.5. Summary of basic parameters for the major communities found in the bidder network (annotated as in Fig. 7.7). N denotes the number of bidders in the cluster, $\langle k_{in} \rangle$ and $\langle k_{out} \rangle$ the average numbers of neighbors within the cluster and in the rest of the network, respectively. By p_{in} and p_{out} the internal and external link density are denoted, respectively. The average link density in the network is $\langle p \rangle = 1.9 \times 10^{-5}$.

Cluster	N	$\langle k_{in} \rangle$	$\langle k_{out} \rangle$	p_{in}	p_{out}
A	200630	10.2	3.4	5.1E-05	5.0E-06
1	84699	10.3	4.0	1.2E-04	5.0E-06
2	29323	9.0	5.2	3.1E-04	6.0E-06
3	76182	10.1	4.1	1.3E-04	5.0E-06
B	102188	18.6	3.9	1.8E-04	5.0E-06
4	44830	24.6	4.2	5.5E-04	5.0E-06
5	26325	14.2	5.2	5.4E-04	6.0E-06
C	19915	14.1	4.3	7.1E-04	5.0E-06
6	20020	14.5	4.3	7.3E-04	5.0E-06
D	124702	16.5	3.8	1.3E-04	5.0E-06
7	74913	17.2	4.1	2.3E-04	5.0E-06
8	41359	16.8	5.9	4.1E-04	7.0E-06
E	183313	15.4	4.2	8.4E-05	6.0E-06
9	73722	13.4	6.5	1.8E-04	8.0E-06
10	47937	17.5	5.9	3.7E-04	7.0E-06
F	74657	10.5	4.9	1.4E-04	6.0E-06
11	62115	11.1	5.0	1.8E-04	6.0E-06
G	31337	11.0	6.0	3.5E-04	7.0E-06
12	18835	11.8	6.1	6.3E-04	7.0E-06
H	19620	10.0	4.4	5.1E-04	5.0E-06
13	18286	9.9	4.4	5.4E-04	5.0E-06

If the network possesses a hierarchical structure in the sense that the clusters obtained at higher values of γ lie completely within those obtained at lower values of γ, then the consensus ordering would not differ from the ordering at $\gamma = 1$. If, however, clusters at lower values of γ overlap and this overlap forms its proper cluster at higher values of γ, the network is not entirely hierarchical. These aspects will become immediately clear by looking at the middle part of Fig. 7.7. For clarity, the borders of the $\gamma = 0.5$ clustering are marked. Clusters 1 and 2 fall entirely within cluster A giving an example of a cluster hierarchy. Cluster 3, however, is split by the consensus ordering into one part A3 belonging to A and B3 belonging to B (see arrows in figure). It is now clear that clusters A and B actually have some overlap which was not visible in the $\gamma = 0.5$ ordering. This overlap is concentrated in cluster 3, parts of which belong stronger to either A or B. Clusters 4 and 5 then fall again completely within cluster B. Clusters C and 6 are practically identical. Cluster D has a number of sub-clusters, the largest of which is 7 and overlaps with cluster E through cluster 8 as before (see arrows again). Group E has two more sub groups 9 and 10 while clusters 11, 12 and 13 fall entirely within clusters F, G and H, respectively. More details about hierarchical and overlapping cluster structures including some toy examples can be found in Chap. 4. Note that in principle, we could have tried to find the optimally fitting block model directly by optimizing (3.15) but did not do so here, because we were specifically interested in finding an optimal assignment into diagonal blocks.

7.2.4.2 Cluster Validation, Interpretation and Time Development

To validate the statistical significance and to rule out the possibility the observed cluster structure is merely a product of the clustering algorithm or the particular method of constructing the network from overlapping cliques of bidders, let us compare the results to those obtained for appropriate random null models. Maximizing Q also for the RNM version of the bidder network, again taking into account those bidders which took part in at least two auctions, a value of $Q = 0.28$ at $\gamma = 1$ is found, which is significantly less than the value of $Q = 0.64$ for the empirical data. Furthermore, the RNM shows all equal-sized clusters, while the real network clearly possesses major and Minor clusters. A random graph with the same number of nodes and links, i.e., disregarding the scale-free degree distribution and the affiliation network structure of the graph, would yield only $Q = 0.23$ (5.65).

Until now we only found groups of bidders which had an increased probability to meet other members of their groups in the auctions they took part in. The eBay product categories are now used in order to find an *interpretation* of the common interests that lead to the emergence of the cluster structure of the bidder network. Since cluster sizes vary and the number of articles in the individual categories is very diverse, the risk ratios (RR) for bidding in one of the 32 main categories are calculated. This risk ratio is defined as

$$RR_{Cs} = \frac{P(\text{bidding in } C | \text{member of cluster } s)}{P(\text{bidding in } C | \text{not member of cluster } s)}, \qquad (7.11)$$

i.e., the ratio of the chance or "risk" of bidding in category C, given a bidder is a member of group s vs. the risk of bidding in category C given the bidder is a member of any group $r \neq s$. Figure 7.8 shows a graphical representation of the risk ratios for clusters A through H and most of the clusters 1 through 13. All spectra are normalized. The exact numerical values can be found in Table 7.6. Clusters from the $\gamma = 1$ assignment are more specific with less entries in the category spectrum and larger RRs.

Cluster A unites bidders interested in articles listed in the baby, beauty, fashion, books, movies and music category. Cluster 1 then represents a more specifically content-oriented user group mainly interested in books, movies and music. As has been shown, cluster 1 is an almost complete sub-cluster of A. Cluster 2 is also a complete sub-cluster of A and encompasses bidders mainly interested in cosmetics and fashion.

Cluster B contains two sub-clusters 4 and 5, both annotated in the toy model category. Closer inspection, however, reveals that cluster 4 is mainly characterized by its interest in model railways while the bidders in cluster

Table 7.6. Risk ratios for *bidding* in one of the 32 main categories during the pre-Christmas season 2004. Shown are only values above 1 signifying an increased interest in articles from this category. Values larger than or equal to 5 are set in bold font. Note how the clusters from the $\gamma = 1$ clustering are more specific than those from the $\gamma = 0.5$ clustering, e.g., there are less categories with an RR larger than 1 and which deviate strongly from 1. Also note how the overlap between clusters A and B is mediated via the toy category and the overlap between clusters D and E via the photo category. For the six largest clusters of the $\gamma = 1$ clustering, the RR of *buying* from the corresponding category during the summer 2005 are shown in parenthesis. See text for further details.

Category	A	1	2	3	B	4	5	C	D	7	8	E	9	10	F	11	G	H
Music	5.6	8.6 (12.0)						1.5										
Books	2.7	3.4 (2.5)		(1.1)														
Movies	12.1	22.1 (9.2)	29.3															
Beauty	5.6	(1.5)	41.0 (20.0)	(1.3)														
Baby	13.7		14.2 (4.9)		5.0	(1.6)	1.1											
Toys	1.3				70.2	21.9 (28.0)	10.2											
Toy Models											1.2	3.4	2.1 (3.9)		1.0	1.4 (2.3)		
Audio		(1.3)						16.2	6.6				(2.2)					
Instruments		(1.5)			1.2	1.8 (1.1)		144.3										
Antiques									88.6	6.3 (3.7)								
Stamps										118.0 (18.0)								
Coins									66.7	88.3 (18.0)								
Collectibles									1.6	1.7 (2.6)								
Jewelry	1.1		1.3						4.5	5.3 (3.1)								
Photo						(1.2)			4.4		47.9	2.8	1.5 (3.2)	3.1		(1.4)		
Computer												19.6	14.7 (12.0)			(1.4)		
Cons. Electr.												6.7	16.1 (4.3)	2.5 (5.9)	2.0			
Mobile												25.2	1.3 (4.7)	33.0		(2.1)		
Games				(1.1)								3.3	2.3 (3.4)	1.4		(1.5)		
Software											1.5	5.1	2.7 (4.0)					
Business													1.1		21.5	16.9 (15.0)		
DIY													2.2		17.9	19.1 (7.0)		
Motors							1.4					1.5	1.6		8.6	8.2 (8.9)		
Travel										(1.7)							58.2	
Tickets				(1.7)									1.7				42.9	
Sports				(1.3)									2.0				1.4	16.0
Office												2.3	1.1 (2.4)	1.6	1.9	(1.9)		
Deli Food													1.5			(2.0)	9.0	
Household	1.9											3.7	3.3 (4.1)		1.1	(1.8)	1.8	
Fashion		(1.0)	2.7	1.7 (4.1)					1.3				1.2				1.1	
Furniture		(1.2)	1.1	(1.6)									1.6			(1.1)	4.2	
Animals		(1.2)	1.0	(1.3)								2.0	1.9			(1.4)		2.3

5 have a passion for model cars, radio-controlled models, slot cars and the like. Note the advantage of clustering based on single articles. The clusters found with one simple unbiased method combine top level categories as in the case of cluster 1 or can only be described by resorting to sub-categories as in the case of clusters 4 and 5. From Fig. 7.7, it had been observed that cluster 3 is responsible for a large part of the overlap between clusters A and B. One can see that users in this group 3 have their main interests in the baby and toy category. The overlap of cluster A and B is hence mediated via the toy category. Members of cluster A and B mainly meet in toy auctions. The interpretation of the other clusters is then equally straightforward.

Bidders in clusters C and the practically identical cluster 6 take interest in audio equipment and instruments. Cluster D represents bidders with an inclination to collecting, their bids being placed in the antiques, jewelry, stamps and coins category (cluster 7). The bidders in cluster E are mainly shopping for technological gadgets, computers, consumer electronics, software, mobile phones, PDAs, etc. (clusters 9 and 10). Their overlapping interest with bidders from cluster D is in items from the photo category (cluster 8).

In groups F and 11, one finds predominantly practically oriented users who place their bids mainly in the categories of automotive spare parts, business and industry (where a lot of tools and machinery are auctioned) and do-it-yourself. Finally, in groups G and 12 one finds event-oriented customers with strong bidding activity in the tickets and travel category and in group H and 13, people bidding on sports equipment are found.

Let us now focus on the time development of the user interests. The data for this analysis were collected during a relatively short time span only (25 days) and the results are based on an extremely sparse data set. Remember that every bidder in the network took part in only three auctions on average. Is it really possible to predict meaningful patterns of consumer interest from such sparse data? One could further argue that the few most active bidders account for a large portion of the bids, thus holding the network together and "defining" the clusters of interest, because they also contribute a large number of links. In order to address this question, the data set was revisited in the beginning of September 2005. From the 6 largest clusters of the $\gamma = 1$ ordering, $10,000$ users each were sampled uniformly and randomly. Note that this removes possible bias toward very active users, they are now represented in the data according to their proportion in the population. The trading history of these users was analyzed as far back as eBay permits – 90 days thus covering the period between June and September 2005. For these $60,000$ users, the product categories of the articles they had bought between June and September were determined.

Again, the risk ratios were calculated, the time of *buying*, i.e., winning an auction, from a particular category and with the new sample of users as basic population. The results are shown in Fig. 7.8 with a dashed background and the cluster id from which the users were sampled in parenthesis. The stability of the interest profiles is quite remarkable. The main interests have remained unchanged as compared to the initial study though in some cases the spectrum has become more diverse. For instance the content-oriented bidders of cluster 1 now also show increased buying activity in the PC games and tickets category. At the same time the main interest has shifted from movies to music. The largest number of product categories with increased chance of bidding in this category is found for cluster 9, the members of which are the most technology affine users anyway and which would be expected to satisfy a very broad range of consumer needs from online vendors. The members of cluster 7 (the collectors) and cluster 4 (the toy model builders) are much more conservative and almost do not change their profile at all. Without secondhand data about the age structure of the bidders classified, one can only speculate that these clusters are formed by older customers who tend to stick to particular categories.

7.3 Summary

In this chapter, we have seen two applications of our block modeling approach, both for the case of finding an optimal block model as well as fitting a diagonal block model to the network. We worked with both weighted and extremely sparse data sets. The analysis did not need any prior knowledge or the definition of any kind of similarity measure between the nodes in the network in order to work. Rather, secondhand information such as the geographical location for the countries in the world trade data or the taxonomic information about articles provided by eBay was used solely to interpret our results.

With regard to the eBay study it is interesting to note that one can classify 85% of all users into only a fistful of well-separated, large groups, all of which have a distinct profile of only a few main interests as revealed by annotating the articles in the taxonomy of product categories. Some of the clusters show sub-clusters or overlap with other clusters. The interest profiles identified are remarkably stable. Sampling randomly from the clusters and checking what these users bought during a 3-month period in the summer of 2005, one finds that the profiles of articles bought were almost identical to those from the classification 6 months earlier.

This is striking because virtually everything is offered on eBay and one would expect users to satisfy a much broader range of shopping interests. However, it appears that the major clusters mainly correspond to people's favorite spare time activities. The apparent stability of user's buying and bidding behavior may reflect the permanence of their interests, which is also stabilized by their social environment and activities. The clear signature in the market data may stem from the fact that users tend to buy online only articles where they have some experience and expertise in. Users seem hesitant to bid on articles from categories in which they have not previously bid in.

This may be due to the fact that inexperienced users cannot judge what is a fair price for an article in an auction and they have difficulty assessing to what extent the article offered really suits their needs. At the same time, user's interests are reinforced by online recommender systems [37, 38], which suggest similar articles to those already bought by the user. This temporal stability corroborates the hypothesis that the presence of latent interest profiles in the society per se leads to the emergence of user groups with common interest. Transparent markets such as online auction sites in which users act independent and anonymously are perfect starting points for research into this collective behavior. The methods presented in this monograph provide the tools to start this endeavor.

References

1. M. Mahutga. The persistence of structural inequality? a network analysis of international trade, 1965–2000. *Social Forces*, 84(4):1863–89, 2006.
2. R. Nemeth and D. Smith. An empirical analysis of commodity exchange in the international economy: 1965–1980. *International Studies Quarterly*, 32(3), 1988.
3. D. A. Smith and D. R. White. Structure and dynamics of the global economy: Network analysis of international trade 1965–1980. *Social Forces*, 70(4):857–893, 1992.
4. M. Rosvall and C. T. Bergstrom. An information-theoretic framework for resolving community structure in complex networks. *Proceedings of the National Academy of Sciences of the United States of America*, 104(18):7327–7331, 2007.
5. S. Wasserman and K. Faust. *Social Network Analysis*. Cambridge University Press, New York, 1994.
6. S. Wasserman and C. Anderson. Stochastic a posteriori block-models: construction and assessment. *Social Networks*, 9:1–36, 1987.
7. M. E. J. Newman and E. A. Leicht. Mixture models and exploratory data analysis in networks. *Proceedings of the National Academy of Sciences of the United States of America*, 104(23):9564–9569, 2007.
8. C. Kemp, T. L. Griffiths, and J. B. Tenenbaum. *Discovering Latent Classes in Relational Data*, pp. 2004–2019. MIT AI Memo, Cambridge, 2004.
9. C. Kemp, J.B. Tenenbaum, T.L. Griffiths, T. Yamada, and N. Ueda. Learning systems of concepts with an infinite relational model. In *The Proceedings of the Twenty-First National Conference on Artificial Intelligence (AAAI '06)*, Boston, MA, 2006.
10. K. Nowicki and T.A.B. Snijders. Estimation and prediction for stochastic block-structures. *Journal of the American Statistical Association*, 96(455):1077–1087, 2001.
11. Special report on ebay. *The Economist*, 375(8430), 06/2005.
12. I. Yang, H. Jeong, B. Kahng, and A. -L. Barabási. Emerging behavior in electronic bidding. *Physical Review E*, 68:016102, 2003.
13. A. Ockenfels and A. E. Roth. Late and multiple bidding in second price internet auctions: Theory and evidence concerning different rules for ending an auction. *The American Economic Review*, 92:1093, 2002.
14. I. Yang and B. Kahng. Bidding process in online auctions and winning strategy: rate equation approach. *physics/0511073*, 2005.

15. M. E. J. Newman. Mixing patterns in networks. *Physical Review E*, 67:026126, 2003.

16. A. K. Jain, M. N. Murty, and P. J. Flynn. Data clustering: A review. *ACM Computing Surveys*, 31(3):264–323, 1999.

17. P. Arabie and L. J. Hubert. Combinatorial data analysis. *Annual Review of Psychology*, 43:169–203, 1992.

18. M. Steinbach, L. Ertöz, and V. Kumar. *New Vistas in Statistical Physics – Applications in Econo-physics, Bioinformatics, and Pattern Recognition*, chapter Challenges of clustering high dimensional data. Springer-Verlag, Berlin Heidelberg, 2003.

19. R. Bellman. *Adaptive Control Processes: A Guided Tour*. Princeton University Press, Princeton, 1961.

20. K. Beyer, J. Goldstein, R. Ramakrishnan, and U. Shaft. When is 'nearest neighbor' meaningful? In *Proceedings of 7th International Conference on Database Theory (ICDT-1999)*, pp. 217–235, Jerusalem, Israel, 1999.

21. M. J. Greenacre. *Correspondence Analysis in Practice*. Academic Press, London, 1993.

22. R. Agrawak and R. Skrikant. Fast algorithms for mining association rules. In 20th VLDB Conference, pp. 487–499, Santiago, Chile, 1994.

23. J. Hipp, U. Güntzer, and G. Nakhaeizadeh. Algorithms for association rule mining – a general survey and comparison. *SIGKDD Explorations Newsletters*, 293(1):58–64, 2000.

24. E. -H. Han, G. Karypis, V. Kumar, and B. Mobasher. Hypergraph based clustering in high-dimensional data sets: A summary of results. *Data Engineering Bulletin*, 21(1):15–22, 1998.

25. B. Sarwar, G. Karypis, J. Konstan, and J. Riedl. Item-based collaborative filtering recommendation algorithms. In Proceedings of *WWW10*, pp. 285–295, Hong Kong, 2001.

26. M. L. Goldstein, S. A. Morris, and G. G. Yen. Problems with fitting to the power-law distribution. *European Physical Journal B: Condensed Matter and Complex Systems*, 410:255–258, 2004.

27. M. E. J. Newman. Power laws, Pareto distributions and Zipf's law. *Contemporary Physics*, 46:323–351, 2005.

28. R. L. Axtell. Zipf distribution of u.s. firm sizes. *Science*, 293:1818, 2001.

29. S. H. Strogatz. Exploring complex networks. *Nature*, 410:268–276, 2001.

30. M. E. J. Newman. Properties of highly clustered networks. *Physical Review E.*, 68:026121, 2003.

31. M. E. J. Newman. Random graphs with arbitrary degree distributions and their applications. *Physical Review E*, 64:026118, 2001.

32. M. E. J. Newman, D. J. Watts, and S. H. Strogatz. Random graph models of social networks. *Proceedings of the National Academy of Sciences of the United States of America*, 99:2566–2572, 2002.

33. E. Ravasz, A. Somera, D. A. Mongru, Z. N. Oltvai, and A. -L. Barabási. Hierarchical organization of modularity in metabolic networks. *Science*, 297:1551, August 2002.

34. M. Girvan and M. E. J. Newman. Community structure in social and biological networks. *Proceedings of the National Academy of Sciences of the United States of America*, 99(12):7821–7826, 2002.

35. M. E. J. Newman. Modularity and community structure in networks. *Proceedings of the National Academy of Sciences of the United States of America*, 103(23):8577–8582, 2006.

36. L. Danon, J. Dutch, A. Arenas, and A. Diaz-Guilera. Comparing community structure indentification. *Journal of Statistical Mechanics*, P09008, 2005.

37. P. Resnick and H. R. Varian. Recommender systems. *Communication of the ACM*, 40(3): 56–58, 1997.

38. J. A. Konstan, B. N. Miller, D. Maltz, J. L. Herlocker, L. R. Gordon, and J. Riedl. Grouplens: Applying collaborative filtering to usenet news. *Communication of the ACM*, 40,(3): 77–87, 1997.

8

Conclusion and Outlook

This monograph has dealt with exploratory data analysis in relational data, specifically with the detection of common patterns in the link structure of networks. Such connectivity patterns lead to a blocking structure in the adjacency matrices of networks when ordering rows and columns such that rows and columns corresponding to nodes with the same pattern of connectivity are adjacent. As such, one can understand the analysis as a kind of dimensionality reduction for sparse, relational data sets.

There were two primary goals to be achieved. First, to develop a simple but flexible method which would allow us to deal with very large and very sparse data sets to cope with the ever increasing amount of empirical data that is collected by scientists across the disciplines using novel experimental techniques and information technology. Dimensionality reduction is usually part of the early stages of exploratory data analysis and results are used to interpret the data and decide on further analysis steps or experiment planning. The second and equally important goal is therefore to guard against the deception of randomness, i.e., making sure we do not mistakenly regard spurious structures due to random fluctuations in the data as true patterns.

The approach taken in this monograph is mapping the problem of pattern detection onto an optimization problem. Starting from a very general *ansatz*, we obtained a novel quality function for the fit between the original high dimensional network and its low dimensional representation, the so-called image graph or block model. The general philosophy of this quality function is to find the assignment of nodes into blocks which deviates maximally from a given, generally random, null model for the connections in the network. Hence, pattern detection is understood as detecting deviations from null models. In being able to deal with arbitrary null models lies the flexibility of the quality function. The random null model for the network may be given as prior information or inferred directly from the data.

For two specific random null models, computationally efficient local update rules were given which allow us to find optimal image graphs using only the sparse connections in the networks and a global bookkeeping term. This allows

Reichardt, J.: *Conclusion and Outlook*. Lect. Notes Phys. **766**, 149–151 (2009)
DOI 10.1007/978-3-540-87833-9_8   © Springer-Verlag Berlin Heidelberg 2009

for easy parallelization of the optimization routines as well as for the analysis of very large systems. Further research into such null models could lead to, e.g., expressions suitable for Barabási–Albert networks where the link probability depends on the age of the nodes or for k-partite networks or expressions which allow to take degree correlations into account.

An important special case is when the image graph of a network consists of isolated nodes with self-links only. Fitting such an image graph to a network using the proposed quality function then means to partition the network into a number of cohesive subgroups of nodes, so-called modules, communities or clusters, which are densely connected within, but only sparsely between groups. Then, the proposed quality function is called modularity and bears a formal analogy with the energy of an infinite range q-state Potts spin glass. This analogy allows us to use the toolbox of statistical mechanics to obtain expectation values for the quality function in cases of purely random networks. Hence, we are able to set a threshold to discern true structure in the data. Only when the quality of the fit exceeds the expectation value for a purely random network, we can be sure we have found true structure in the data.

For entropic reasons, when clustering random graphs, clusters of equal size are found. This means that in purely random networks clustering and partitioning into equal-sized parts are equivalent, i.e., minimum cut and maximum modularity are then equivalent problems. For random graphs with arbitrary degree distributions expectation values for the modularity were calculated using the replica method and the cavity method. For a partition into only two clusters the expected modularity scales as $\propto \langle k^{1/2} \rangle / \langle k \rangle^{1/2}$. The sparser a random graph, i.e., the smaller the average degree, the higher the expectation value for the modularity found in this graph. Since structure is only detectable if it leads to a fit score which exceeds the expectation value for a purely random graph, we find that community detection is more difficult in sparser networks which show higher values of intrinsic modularity. Together with our findings on the expectation value of modularity for different degree distributions, we see that community detection is simpler in networks with a broader degree distribution.

The next question is naturally by how much the fit score has to exceed a random expectation value so that one can expect a certain accuracy in recovering a cluster structure in the network. Hence, the problem of recovering planted cluster structures from networks using minimum cuts was studied using the cavity method. As expected from the considerations following the replica calculations, one finds a transition from undetectable to detectable cluster structure as the density of intra-cluster links is increased at the expense of inter-cluster links.

Surprisingly, this transition is sharp and a result of the sparsity of the data and not a finite size effect. Even in infinitely large networks with finite average connectivity, this transition exists. Hence, there may exist cluster structure which is not detectable by unsupervised methods in principle. It is hidden between alternative solutions leading to better fit scores, but which are uncorrelated with the structure underlying the data. There exists a detection threshold for cluster structure in networks. Benchmarking showed that our proposed method for detecting cluster structures in networks already reaches this threshold and hence produces near-optimal results. In the future, these calculations will need to be extended to more general image graphs. Besides giving expectation values of the quality of fit of general image graphs to random graphs in the thermodynamic limit, finite size corrections should be applied in order to give an estimate of the variance of the fit score around these expectation values. This is necessary in order to eventually be able to give p-values for block structures found in complex networks.

Two examples from economics show the power of the proposed method for block structure detection and how it can be used efficiently to study problems from a variety of fields.

Questions of the emergence and evolution of the block structures pose further interesting challenges for future research. Are there microscopical models that will produce certain kinds of block structures? Are different types of networks characterized by different types of block structures? The interplay between block structures and dynamical processes on networks is only vaguely understood. Consider for example the question of global structure vs. local information. While we have used knowledge of the entire network in our analysis so far, agents operating in networks are generally only aware of their local neighborhood. Since block structure is a general feature of networks, an interesting question to ask is whether agents can still be aware of global structure despite having only localized interactions. Possible mechanisms could be repeated interactions and diffusion of information in networks. Many having questions about the interplay of topology and dynamics such as this still remain unanswered and pose interesting challenges for complex networks research.

Throughout this monograph, we have successfully applied tools and results of physics that were originally developed for the study of condensed matter phenomena in a completely different context. This use of analogies to physical systems and methods developed in the realm of physics can be a fruitful approach in many fields of science. In this work, the connection between a spin system and a combinatorial optimization problem, the detection of block structure in networks, was exploited. Both a solution method and an understanding of the optimization problem itself could be derived. Naturally, the application of physics methods to diverse non-physical problems will also lead to an advancement of these methods from which physics itself will again benefit. Though this work could only give an account of one cycle of this iteration, its further pursuit is a promising and rewarding endeavor.